Total Nutrition
Feeding Animals for Health and Growth

Clifford A Adams

NOTTINGHAM
University Press

Nottingham University Press
Manor Farm, Main Street, Thrumpton
Nottingham, NG11 0AX, United Kingdom

NOTTINGHAM

First published 2001
© Nottingham University Press

British Library Cataloguing in Publication Data
Total Nutrition: Feeding Animals for Health and Growth
Adams, CA

ISBN 1-897676-94-8

Typeset by Nottingham University Press, Nottingham
Printed and bound by The Cromwell Press, Trowbridge, Wiltshire

TOTAL NUTRITION
Feeding Animals for Health and Growth

CONTENTS

Preface

The successful raising of large numbers of animals for food and the production of large volumes of low cost food for the human population is an undoubted testimony to the value of basic and applied animal sciences. Moreover the intensive raising of animals for food is now being implemented on a worldwide basis in both the developed and developing countries.

This has produced the fortunate social situation, in the developed world, where an adequate and available supply of food of animal origin is taken for granted. It has also resulted in a society where ever-fewer people are actually concerned with, or have any knowledge of, the processes of providing food of animal origin. In addition the various food safety scares in the recent decades have focused consumer and political attention upon these processes of animal production. This has in turn led to stricter demands and controls by legislative authorities, consumer groups and retail organisations.

It is important that those engaged in raising of animals for food appreciate and respond to these various concerns. This was the impetus for my first book; *NUTRICINES Food Components in Health and Nutrition*. Here I attempted to show that there are many natural components of feed and food, the nutricines, that have valuable and beneficial effects in relating health to nutrition.

This second book, *TOTAL NUTRITION*, is an attempt to develop further the use of nutrients and nutricines in animal nutrition. The objective in Total Nutrition is to obtain the maximum value of various nutrients and nutricines at all stages of the animal production chain from feed raw materials to the nutrition of the mature animal. In Total Nutrition animal feeding must be done with feed components that are likely to be acceptable to the modern consumer. Judicious use of various nutrients and nutricines seems to offer a good practical solution to problems of health maintenance, good zootechnical performance, and reductions in adverse environmental impact of intensive animal production.

The book tries to follow a logical sequence. Problems of consumer perception and food safety are discussed in Chapter 1, followed by feed quality and hygiene in Chapter 2. Feed palatability and digestion and absorption of nutrients are discussed in Chapters 3 and 4 respectively. Animal health issues are the focus of Chapters 5, 6 and 7. Here the major topics are management of the gastrointestinal tract, the immune system and non-infectious diseases. Chapter 8 considers new ideas over the monitoring of animal performance and Chapter 9 addresses consumer demands, ethical and environmental issues.

Total Nutrition offers the exciting possibility of the further development of animal nutrition through improved use of a wide range of feed components, both nutrients and nutricines. Feeding animals for health and nutrition must surely be a major objective for future progress in animal science.

Clifford A. Adams

1 Problems of Perception: Animal Production, Food Safety and Public Health

Animal husbandry, to produce meat, milk and eggs as food items, has been an occupation of the human race from time immemorial. Despite our modern sophistication and technological advances, feeding and raising of animals for food remains and always will remain one of the most fundamental aspects of human activity. Food of animal origin plays an important and indispensable role in human life. It is nutritionally of high value and on a social level it is highly desired.

The great social desire for food items of animal origin is vividly manifested in the developing world. It is well recognised that as countries develop economically their populations increase consumption of foods of animal origin. There seems to be a universal desire for large quantities of foods of animal origin as a part of the daily diet. The nature of these foods varies somewhat from country to country depending on religious and cultural differences. For example; cattle, pigs, horses, dogs, snakes and crocodiles are used as meat sources on a global basis but all these species are not eaten in all countries. Poultry meat, eggs and fish seem universally accepted.

It is also important to emphasize that meat, eggs and dairy products are excellent sources of protein for human nutrition and are very valuable food items. One hundred grams of lean meat provides up to half our daily protein requirement (Saucier, 1999). It is also a good source of vitamin B complex and of minerals, especially iron, which is readily bioavailable in meat. Eggs and dairy products are also widely recognised for their high nutritional value and easy digestibility.

Meat, milk and eggs are not only valuable foods in their own right, but they have also led to an enormous food manufacturing industry which is of great economic value. We have available today innumerable ready-prepared foods based on meat. Dairy products lend themselves to the production of hundreds of different foods, including cheeses, butter, cream, yoghurts, and ice creams. Eggs are used in a multitude of baking processes as well as in prepared foods.

Meat, milk and eggs and products derived from these animal foods are a part of the daily diet of many millions of individuals worldwide and this in turn requires the raising of many millions of animals for food use. Animal husbandry is truly a very large and global industry both in terms of volumes of materials used and in terms of numbers of animals raised.

Over the last 25 years worldwide production of manufactured animal feeds has grown almost parallel with the increase in population of some 1.5% annually. In 2000 probably about 620 million tonnes will be produced and this will most likely increase to about 700 million tonnes by 2005. In 1999 Asia was the largest producing area with a volume of around 140 million tonnes and the European Union (EU) was second with a volume of 122 million tonnes (Schumacher, 2000). Animal husbandry is certainly an important economic component of the agricultural activities of most countries.

In the 20th Century the general expansion in scientific knowledge made major contributions to improved animal production systems. This has consequently led to the fortunate modern situation where food of animal origin is widely available in large volumes and at low cost. Certainly in the developed countries, adequate supplies of meat, milk and eggs are taken for granted. Yet all too often the tremendous scientific and technical advances which have supported this food supply are frequently overlooked.

The varied and extensive food supply now available in the developed world is also changing the concepts of human nutrition. Future developments in food science will focus on the components of foods, both the nutrients and nutricines (Adams, 1999), that have the potential to modulate functions in the body such as gene expression or immune status. There is an increasing awareness that many non-infectious diseases such as heart conditions, many cancers, diabetes and joint diseases may be subject to nutritional modulation. The objectives now are to use nutrition to promote well-being and good health and to reduce or avoid diseases (Schneeman, 2000). Modern animal husbandry must also respond to these changes in human nutrition and the concept of "TOTAL NUTRITION" is a useful strategy for animal production where attention is increasingly focused on raising animals with minimal medication and using nutrition to promote health as well as growth.

Consumer perceptions

In the 21st century, animal production although highly productive, faces obstacles and difficulties, both real and perceived. Several factors have contributed to this situation. The widespread availability of food of animal origin, and the general prosperity of the population means that obtaining food is no longer a major concern for the majority of the people in the developed world. Fortunately there is a wide choice of food items available and it is relatively easy for consumers to switch from eating beef to pork or chicken for example if a food safety issue arises. This has been seen with various food safety scares such as salmonella in poultry and BSE in cattle. In both cases massive reductions in purchases by the consumers occurred. It is very difficult for the animal production industry to predict where the next issue will arise.

Increased urbanization and concomitantly a reduced proportion of the population engaged in agriculture means that fewer people are conversant with or knowledgeable about the processes of food production from animal husbandry. Consequently the consumer is in many cases completely in the hands of the mass media for information and knowledge on aspects of food production and food safety. Over recent years the animal production industry also has perhaps not been sufficiently committed to public education on issues of food production and food safety. These socio-economic factors together with various unfortunate food safety scares have focused enormous critical attention on the animal husbandry industry worldwide.

In recent years a succession of major issues have occurred that have undoubtedly moulded public perception. These food scares commenced with the *Salmonella* problems in poultry of the 1970s. This was shortly followed by the appearance of BSE (Bovine Spongiform Encephalopathy) in cattle in the 1980s and its' resurgence in Continental Europe in 2000. There have been increasing concerns expressed by various health professionals about the development of bacteria resistant to antibiotics. The Dioxin crisis in feed fats in 1999 and the introduction of genetically modified crops (GMOs) further increased public concern over various aspects of food safety

Also in the 1980s, animal welfare issues gained widespread publicity. A whole plethora of animal welfare issues have been debated including live animal transport, tail docking in piglets, sow stalls and tethers, beak trimming in poultry, feed restrictions in poultry breeders, battery cage production of eggs, and leg weakness in several species.

These issues have all been given major publicity by the media and unfortunately much of this has generated a negative perception of the animal production industry, especially in the European Union. Consequently both the general public and government officials now direct increased attention to methods of food production and to food safety and this has had a significant impact upon all phases of animal production. It has resulted in greatly increased new legislation concerning animal welfare and nutrition.

In 2000 the European Commission published a white paper setting out the future direction of food production in the EU. This document specifically states that animal feed must be treated under the same conditions as food. This will require much more transparency in systems of animal nutrition and production. It will also require considerations of animal health and welfare as well as productivity.

Events in the end of the 20th Century also encouraged the major retail groups to become much more actively involved in checking and controlling the production of the food products they buy. The major supermarket groups realized that by taking the lead on such issues as animal welfare and food safety the strength and image of their brands would be enhanced. In the UK in particular the supermarkets have become very involved in all stages of animal production. This includes selection of feed ingredients such as antibiotic growth promoters, colorants for egg yolks, meat and bone meal and animal fats. They make decisions on aspects of welfare such as maximum stocking density and transport of live animals. These measures have encouraged the use of more extensive systems such as free-range egg production.

The events in recent years have certainly changed the balance of power between producers of food and the consumer with the supermarkets following a much more proactive policy towards the supplier. In reality the supermarkets can now impose regulations on feed manufacturers and animal producers which are more stringent than those required by governments. This is now presented as a response to consumer concerns over food safety and quality.

There are however dangers here as pressure groups representing a minority interest can now sharply focus media attention on a topic to forces changes in animal production which are not necessarily supported by scientific evidence. One recent example of this is the exclusion of the red carotenoid, canthaxanthin, from layer feed in the UK even though this is accepted by legislative and scientific authorities as posing no risk to human or animal health. Another potentially more damaging example is the attempts by various environmental pressure groups to prevent the distribution of the genetically modified "golden rice " (Potrykus, 2001). This modified rice is a major scientific breakthrough in crop production as it adds an essential dietary component to one of the most important staple foods of the poor and developing world. This genetically modified rice strain contains genes to produces ß-carotene which is a precursor of vitamin A. In many poorly developed countries there is a lack of vitamin A in the diet which leads to blindness in children. Golden rice could be of great benefit here.

There clearly are many consumer concerns over food safety and this has focused attention upon methods of animal feed manufacture and upon systems of raising animals for food. One manifestation of this concern is the significant and growing interest in "organic food" production where crops and animals are produced without the use of pesticides, chemical fertilisers or antibiotics. Detailed comparisons of the nutritional quality of foods produced under organic versus conventional systems are not easy to make. Foods of plant origin produced under organic systems may have a higher content of some nutricines but there seems little difference in overall nutrient content (Brandt and Molgaard, 2001) and much more work will be needed here.

The quantity of organic food production will inevitably be limited by the exigencies of the production system and organic food will always be significantly more expensive than the products from intensive agriculture. Organic farming leads to reductions in crop yields of 30-50% and is not a feasible global option for meeting the food requirements of the world's population (Avery, 1999). Whilst organic foods can supply a niche market for people prepared and able to pay higher prices it can never deliver the large quantity of low cost foods that a modern society needs. Furthermore organic food is no more immune from contamination by pathogenic micro-organisms than food produced under intensive systems.

A whole plethora of events have generated serious problems of consumer perception of modern agriculture (Fraser, 2001). Some of these perceptions are valid and there is certainly an obligation upon the animal production industry to respond to these legitimate concerns. Other perceptions are not scientifically founded and these need to be combated by careful and reasoned arguments. It will be important for the industry to be more transparent and to develop production systems more in tune with consumer requirements. This is where a concept of Total Nutrition may offer some assistance with a focus on using nutrition to support health as well as growth of food-producing animals.

Food safety

Because of the many adverse consumer perceptions about food the European Union has made food safety a top priority (Vanbelle, 2000). The central element of the EU approach to food safety is that it must be based on a global integrated system throughout the food chain from farm to fork across all sectors of the feed, animal production and food manufacturing industries. This means that feed manufacturers, farmers and food manufacturers all have the primary responsibility for food safety.

It is also important to establish precisely what are the main threats to food safety and human health. These are listed in Table 1 and clearly micro-organisms, both bacteria and moulds represent a continuing threat to food safety. These have the potential to be lethal for the human or animal population and must always be controlled. Other aspects such as drug residues and pesticide residues are not usually life-threatening but rather undesirable. Use of food additives is often considered a health risk or food safety issue but judicious application of preservatives and antioxidants also has many positive health benefits as discussed in chapters 2, 5 and 7.

Threat	Example
Naturally occurring bacteria	Listeria in fruits and vegetables
Bacteria from intestines of animals contaminating food	Salmonella and Campylobacter in meat
Naturally occurring toxic substances	Algal toxins in shellfish, mycotoxins on fruits and cereals
Residues from medical treatment of animals	Antibiotics
Environmental contaminants	Dioxins, heavy metals
Pesticide residues	In fruits and vegetables
Food additives	Flavours, colours, preservatives

The safety of foods derived from animal production can be influenced in many ways, including the presence of chemical residues, such as pesticides and antibiotics. Perhaps the most important food safety factor is the possible contamination by pathogenic bacteria species such as; *Bacillus, Campylobacter, Clostridia, E. coli, Listeria, Salmonella, Shigehella, Staphylococcus* and *Yersinia* (Jackson, 1990). Bacteria are in a constant state of evolution with infinite capacities to respond to our efforts to control them. Consequently food safety is and always will be a perennial problem and will most likely never be completely solved. This does not imply however that strenuous efforts should not be made to improve food safety. Indeed current events require nutritionists, feed manufactures and animal producers to develop new and novel strategies in response to these concerns.

The food safety scares and issues of antibiotics and animal welfare have generated a whole plethora of new demands and of potential solutions (Knudsen, 2001). There is an increasing tendency to require the raising of food animals without recourse to antibiotics and other drugs. There is a certain body of public opinion that equates everything natural with high levels of safety and quality and much discussion of sustainable agriculture. This attitude overlooks the fact that many compounds of natural origin are toxic and that extremely large quantities of food of animal origin are required to feed the human population. Furthermore in many cases arguments promoting sustainable agriculture are programmes to promote sustainable poverty. Nevertheless practices of modern animal husbandry and animal nutrition have a very high profile and are inevitably visibly linked to food safety. There are now new criteria for safety when it comes to the feeding, healthcare and general treatment of livestock, and these must be taken into consideration in future programmes of animal production. There is an increasing onus on the animal production industry to convince sceptical consumers and regulatory officials that food products of animal origin are produced in the best possible way and that food safety is a major issue.

Antibiotics

One well-publicised major concern of recent years has been the use of antibiotics as growth promoters in monogastric animals and the generation of resistant strains of bacteria. This practice has certainly been very poorly perceived by the consumer and there is a lot of consumer pressure not to use antibiotics in raising of animals for food.

The growth promoting effects of subtherapeutic levels of antibiotics in animal feeds was discovered in the late 1940s and has subsequently become an important part of modern animal nutrition. This widespread application of antibiotic growth promoters has also been supported by a great deal of scientific research. A comprehensive literature review indicated that in 12, 153 reported trials the addition of antibiotic growth promoters to feeds increased animal performance 72% of the time (Rosen, 1995). With such a large body of supporting data it is hardly surprising that antibiotic growth promoters became standard ingredients in pig and poultry feeds over the last 50 years.

However, concerns about the use of antibiotics in animal production, and the appearance of bacterial resistance were also expressed quite some time ago and is certainly not a new phenomenon (Anderson, 1965). The emergence of ampicillin-resistant *S. typhimurium* strains was traced to the use of this antibiotic to treat or prevent infections in calves. This raised concerns in 1965 that resistant strains of *S.typhimurium* found in calves could transfer resistance to other bacteria such as *E. coli* and subsequently to other human pathogens.

Since that time several other examples of the generation of bacterial resistance to antibiotics have been reported (Witte, 2000). Special attention was paid to avoparcin, virginamycin and tylosin because of possible cross-resistance against therapeutic antibiotics used in human medicine. Perhaps the most serious situation was the isolation of glycopeptide-resistant *Enterococcus faecium* in the intestinal flora of animals and humans (Wegener *et al.,* 1999). The antibiotic, avoparcin is also a glycopeptide and consequently animal husbandry came under suspicion as a reservoir of resistant bacteria. Some strains of *Salmonella typhimurium* DT104 have been shown to have resistance against several antibiotics including ampicillin, chloramphenicol, streptomycin, sulfamethoxazole, and tetracycline. (Bower and Daeschel, 1999). This multi-resistant strain can thrive in an animal being fed sub-therapeutic doses of antibiotic when competing bacteria from the normal flora are suppressed. The resulting reservoir of infectious organisms can persist and be transmitted to other animals and eventually enter the food chain.

It is highly likely that most of the antibiotic resistance problems in human medicine stem from overuse or inadequate control in medical practice for humans. Nevertheless there is concern that enteric bacteria such as *E.coli*, *Salmonella*, and *Campylobacter* with antibiotic resistance can transfer from animals via the food chain or by direct contact. This could establish a population of resistant genes in both animals and humans (Barton, 2000).

Antibiotics can be used in three ways in animal production: therapeutic to treat sick animals, prophylactic to prevent infection in animals and as growth promoters to improve feed utilization and productivity. Use of antibiotics in therapeutic treatment to cure sick animals is usually under the control of veterinarians and is also a welfare issue. Use of antibiotics as prophylactics or for growth promotion is much more of an economic tool to obtain more efficient production and it is in these areas that alternative systems of animal production must be applied.

Widespread publication of instances of bacterial resistance together with other food safety issues culminated in banning of many antibiotics from use in animal nutrition in the EU. In general there is now great opposition to the use of antibiotic growth promoters in animal production by consumers and retail organisations.

Opposition to the use of antibiotic growth promoters in animal production however is not new and started in Sweden in 1985 when a new feed law was promulgated. Under those regulations, antibiotic and chemotherapeutic substances were allowed in feeds only on prescription by a veterinarian. This was slightly amended in 1993 when the use of zinc oxide was approved at 2000 ppm in pig starter feeds for two weeks after weaning.

A culmination of all these various trends and concerns was that in 1999 a whole range of pharmaceutical products were prohibited for use in animal nutrition in the EU (Table 2). The only antibiotic growth promoters still permitted are; avilamycin, flavomycin, salinomycin and monensin.

Consumers are concerned about the wholesomeness and safety of food and therefore the animal feed and production industries must be able to convince consumers that they are producing food of animal origin that is both safe and of good nutritional quality. This also means food products or animal origin that have no residues of antiobiotics and with only minimum levels of bacteria that have acquired antibiotic resistance.

<table>
<tr><td rowspan="4">

Table 2
Recent history of antibiotic growth promoters in the EU
</td></tr>
</table>

April 1997	Avoparcin prohibited for use in animal nutrition.
July 1999	Virginiamycin, spiramycin, tylosin phosphate and zinc bacitracin prohibited from use in animal nutrition.
September 1999	Carbadox and olaquindox prohibited from use in animal nutrition.
October 1999	Dinitolmide, ipronidazol and arpinocide prohibited for use as coccidiostats.

Intensive animal production

We must inevitably rely upon intensive modern systems of animal production for the large bulk of our food supplies. Nevertheless there are many radical changes occurring in modern animal husbandry and the situation is by no means static. The animal production industry is now rapidly evolving into an integrated food industry for the 21st Century. It is widely recognised that there is a continuous chain from primary animal feed raw materials to health through nutrition of the human population. It will always be necessary to achieve a satisfactory balance between maximising production volumes, minimising costs and ensuring food safety and customer satisfaction. For many years animal nutrition and animal feed production were hidden links in the overall food chain. Nowadays the animal production industry must be visible, and seen to be engaged in providing consumers with, safe, healthy, abundant and affordable food.

Modern systems of animal production have been extremely successful in delivering large volumes of low cost food to the human population. This requires genetic strains or lines of animals which have the potential for rapid growth and high production rates of the food item of interest, whether it be milk, eggs or meat. Decades of research into animal nutrition, genetics and husbandry permit us to control with great precision the development of our food animals. Factors such as reproductive rates, health status, growth rates, percentage of body fat, shelf life of meat, colour of meat and eggs are all subject to nutritional control.

The demands of a modern society for food of animal origin also requires the raising of large numbers of animals in relatively small areas. This combination of high productivity of animals and large numbers, inevitably mean that animals are exposed to considerable stress during their productive period.

This stress has several origins. The time immediately after hatching or birth is a period of stress for most animals. The gastrointestinal tract of new-born animals is immature and sterile and begins to develop its function and its microflora when it begins to ingest feed. At this time the animal is very susceptible to pathogenic micro-organisms as the animal usually has little or no natural defences. In mammals, particularly calves and piglets, weaning imposes a severe stress when they are frequently moved to another location and receive a new and completely different diet than the milk-based diet they obtained from the mother.

Animals raised for food must obtain adequate levels of nutrients to support rapid growth rates and high productivity. Therefore they must receive adequate quantities of good quality feed to avoid stress. This feed must have the correct nutrient balance and be available to all animals. Feed must be produced as cheaply as possible from available raw materials. Some feed ingredients, however, such as wheat, barley and fats are known to cause digestive stress in some species. Feed inevitably contains micro-organisms and other components which may be toxic and put additional stress on the animal, either through diseases or activation of the immune system. There is constant pressure of infectious diseases which can rapidly spread throughout populations kept in close proximity to each other. Non-infectious, or metabolic diseases such as ascites in poultry and lameness in many species causes stress. Extensive preventive medication and vaccination is a major contributor to stress.

Modern animal production has traditionally dealt with some of the problems of stress by use of antibiotics both in therapeutic and in sub-therapeutic quantities. However the widespread use of antibiotics is no longer tenable and alternative systems to overcome stress and to maintain efficient animal production must be sought. The links between disease, health, nutrition and environment all need to considered. Now the focus must be directed to understanding what keeps an animal healthy and how to avoid disease without the use of pharmaceutical products.

Despite the impressive body of scientific literature testifying to the efficacy of antibiotic growth promoters it was also recognised that these antibiotics gave greater growth performance enhancement in environments with poor hygiene compared to those with good hygiene status. It was also established that chicks raised in a germ-free environment showed better growth rates than chicks raised in a conventional environment and such chicks did not respond to subtherapeutic levels of antibiotics as growth promoters. Clearly there is an interaction between environment and animal growth performance. These observations suggest that the presence of various microbial pathogens in the normal environments were responsible for impaired growth of commercial animals.

Effective response to these demands requires new strategies for animal nutrition and perhaps a new concept known as "TOTAL NUTRITION" where health maintenance, disease avoidance, general nutrition and environmental impact, are all considered as requirements of feeds.

Total Nutrition

Animal agriculture, as with other human activities, shows a progression over time (Schneeman, 2000). Initially there is the problem of quantity of production, of supplying adequate amounts of food for the human population and this has been largely achieved in the developed world. There is always a continuing challenge to improve the efficiency of production and this usually results in lowered costs. Food costs in relation to buying power are certainly lower in the developed world in the 21st Century than they have ever been. In the UK food costs have consistently declined as a proportion of household expenditure over the last 30 years and in 2000 it was about 10% of expenditure (Thomas, 2001).

However once adequate supplies of low-cost food are available concerns shift to issues of food safety and environmental impact and to the relationship between health and nutrition. It now becomes possible to search for ways to modify and improve the food supply to promote health and to avoid disease. This focuses more attention upon the mechanisms of animal production where also nutrition has to become more closely linked to animal health. This becomes even more significant when there is strong sociological pressure to reduce use of drugs and medicines in production of animals for food.

Total Nutrition is a strategy with which to approach production of animals for food in the 21st Century and it must encompass several difficult tasks. Food of animal origin must be produced in large quantities and as cheaply as possible. These food products must be absolutely safe for the human consumer. Minimal amounts of antibiotics and other medicinal products should be used in the raising of animals for food. Nevertheless the animals should be raised under conditions of good welfare and be kept free from diseases. The effects of maintaining large numbers of animals in relatively small areas should not generate environmental pollution. Such are the challenges for Total Nutrition in the present climate.

On a more detailed level, Total Nutrition is concerned with the whole feed chain from quality of raw materials to control of metabolism within the body of the animal to the final human food items. This now requires design of feeding programmes to encompass many different functions starting with safety of feed raw materials and of stored feed. It must influence the acceptance and consumption of feed, the digestion and absorption of nutrients, and modify the microflora in the gastrointestinal tract. It must support the immune system and avoid oxidative stress. Feed formulations and animal production should generate the minimal environmental nuisance and pollution.

To achieve Total Nutrition animal feeds have to be approached in a slightly different manner than in the past when feed was usually considered only a source of nutrients. In practical reality however, animals consume a great diversity of different molecules in feed in addition to the conventional nutrients. Many of these molecules have been regularly consumed for thousands of years and play an important role in animal and human health and welfare even though they may not have been deliberately used or recognised.

The entire range of different molecules found in feeds can be classified into two major groups; NUTRIENTS and NUTRICINES (Adams, 1999). Nutrients are the generally recognised components of feed such as carbohydrates, proteins, fats, minerals and vitamins. Nutricines are components of feeds that exert a beneficial effect upon health and metabolism, yet are not direct nutrients (Table 3). Important nutricines are; antioxidants, emulsifiers, enzymes, flavours and colours, non-digestible oligosaccharides and organic acids. The nutricines are those components of feeds that link health and nutrition and will play an increasingly important role in Total Nutrition.

Nutrients	Nutricines
Carbohydrates	Antioxidants
Fats	Colours
Minerals	Emulsifiers
Proteins	Enzymes
Vitamins	Flavours
	Non-digestible oligosaccharides
	Organic acids

Table 3
Total feed components consist of nutrients and nutricines

In Total Nutrition feed has to have multiple properties encompassing nutritional, health and environmental concerns. This has to be accomplished using naturally occurring compounds i.e. nutrients and nutricines. Thus the feed becomes functional in its nature and confers both nutritional and health benefits to the animal consuming it.

A major task in Total Nutrition is to understand the links between diet, health, disease and environment. This is a very complex system, full of paradoxes. For example animals live in a hostile environment where they must compete against a host of other living organisms on this planet which includes plants, insects, fungi, bacteria and viruses. These hostile environmental factors must be controlled and overcome to obtain efficient animal production. At the same time animals consume feed produced in the environment and they produce in turn various waste products such as methane gas, urine and faeces which are potential serious environmental pollutants.

Feed consumption is another paradox. Consumption of feed is clearly essential for life and growth of animals yet it is also a potential route for the entry of pathogenic organisms into the body and represents a danger to the health of the animal. Careful feed manufacture, storage and processing is necessary to minimise contamination by natural pathogens. Nutritious feed, by its very nature, is extremely nutritious for pathogenic organisms, which includes rats, mice, insects, fungi, yeasts and bacteria. All of these organisms can and will contaminate feeds if given an opportunity. Even under good conditions of feed manufacture and hygiene a considerable number of potentially pathogenic micro-organisms will always be consumed in feeds. It is important however to minimise this and to maintain nutritional quality of stored raw materials (Chapter 2).

Feed intake is the second major concern of Total Nutrition since feed that is not eaten, is of no nutritional value to an animal, and therefore all feeds must have acceptable organoleptic characteristics which persist throughout the shelf-life of the feed. Many factors influence feed intake such as environmental and disease stresses which commonly occur in pigs and poultry produced under commercial conditions. Immune stimulated, or disease-stressed pigs and chickens grow slower than animals not stressed. (Baker and Johnson, 1999). This slower growth is due to a reduction in feed intake and a concomitant reduction in protein accretion during the disease challenge. The flavour characteristics of feeds, or the palatability, are also in many cases indicators of feed quality. The flavour of fresh good quality feed is obviously different from that of rancid or moulded feeds. Feed quality, disease and environmental stress all generally cause a reduction in voluntary feed intake of animals which will subsequently affect animal performance (Chapter 3).

Ingested feed ingredients are broken down into the basic essential nutrients by the process of digestion in the gastrointestinal tract. Enzymatic digestion in the gastrointestinal tract is only the first step in feed utilization. Absorption of nutrients from the gastrointestinal tract is extremely important as only those nutrients which have been absorbed, can be used by the animal for growth and production. Absorption is a complex physiological process and is influenced by phospholipids which are members of the lecithin family. Enzyme and phospholipid nutricines play an important role in assisting digestion and absorption of nutrients (Chapter 4).

The gastrointestinal tract is the largest organ in the body. It is a very complex organ and functions as an interface between the metabolism of the body and the environment. It provides an extensive surface area over which direct contact takes place between a wide array of nutrients, nutricines, microorganisms and exogenous toxins. The intestinal epithelium or lining of the gastrointestinal tract must be maintained in a good physical state to prevent the bulk transport of pathogens into the body, but it must also be sufficiently thin to actively transport nutrients. Inside the gastrointestinal tract is, in physiological terms, still outside the body. Only when digested food components have been absorbed across the wall of the gastrointestinal tract are they physiologically inside the body. Management of the gastrointestinal tract is of vital importance in maintaining good animal health and avoiding enteric diseases (Chapter 5).

Animals have also evolved an immune system to deal with pathogenic micro-organisms which either enter the body in food and are located firstly in the gastrointestinal tract or which enter the body through a physical injury or a wound or through the lungs. This remarkable system distinguishes harmful molecules, known as antigens, which include bacteria, viruses and pesticides from innocuous dietary proteins. It reacts against both types of molecules locally. In the case of microbial antigens it allows a full activation of the immune system if the defences of the gastrointestinal tract are breached. However in the case of dietary proteins it is important to suppress this immune response and allow feed proteins to be digested and absorbed from the gastrointestinal tract.

Animals continuously receive a series of pathogenic and non-pathogenic challenges throughout their lives from feed, water and the environment. Bacterial and viral pathogens cause activation of the immune system through the production of cytokines. This has severe metabolic effects upon the animal. The cytokines are hormone-like protein molecules and they cause reduced feed intake, increased body temperature and poorer feed conversion rates, both in poultry and in pigs (Klasing and Johnstone, 1991; Williams *et al.*, 1997a, 1997b). The cytokines released by each microbial challenge diverts nutrients away from growth in order to support the immune system, and this represents a major obstacle to efficient growth performance. Feed is the largest source of antigens and other chemicals encountered by the animal, and it may also contain components capable of suppressing or of stimulating the immune system. Consequently Total Nutrition must take into account the immunological status of feeds and seek to avoid unnecessary stimulation of the immune system (Chapter 6).

Many chemical components of the environment are toxic and in particular animals have to contend with the "Oxygen Paradox." Oxygen is inherently dangerous to the existence of animal life through various oxidation processes and yet at the same time it is essential for life. Control of oxidation of feeds and within the body is of fundamental importance both to maintain nutritional quality of feed and to reduce the risk of various non-infectious diseases (Chapter 7).

The quantities and type of nutrients necessary for growth of animals is well defined. It is however less well understood whether the general nutritional requirements for growth and maintenance of body weight are equally suitable for disease avoidance, control of oxidation, control of feed pathogens, development of the immune system and maintenance of health in the absence of antibiotics. New concepts on definition of levels of nutrients and nutricines must be established.

In the new concept of Total Nutrition, the minimal level of any feed component, nutrient or nutricine that effects the metabolism and gastrointestinal function in a manner beneficial to good health must be considered. The future is no longer to define optimum nutrient intakes, but rather to determine the Total Nutrition necessary to give optimum health and nutrient status in the animal. Total Nutrition can be assessed by measurement of various functional indices that should be directly related to disease mechanisms or to ill health (Chapter 8).

By thoughtful application of the various nutrients and nutricines available for animal nutrition it is possible to get extra value out of basic raw materials, particularly those inedible for direct human consumption. This leads to the concept of the Feed Sparing Effect, which has both sociological as well as economic implications. Total Nutrition will also have a beneficial environmental impact leading to a reduction in environmental pollution (Chapter 9).

The concept of Total Nutrition must address the major concerns of the modern consumer. Total Nutrition requires the judicious application of both nutrients and nutricines so as to obtain highly efficient animal production that is not based on excessive use of antibiotics and other medicinal products. By using the strategies outlined here it will be possible to fulfil the basic requirements of the modern consumer.

Total Nutrition must promote animal nutrition and growth and alleviate stress, at all stages of the food chain, from feed raw materials to food as depicted in Figure 1 (Adams, 2001). The stress is of both metabolic and environmental. Environmental stresses are mainly due to the necessity of raising large numbers of animals in close proximity to each other. This is essential if adequate supplies of food are to be produced for the human population but it nevertheless exposes the animals to high infection pressures and to atmospheric toxins such as ammonia. Vaccinations, which are indispensable to maintain animal health and control infectious disease, add another level of stress to the animals. Metabolic stress due to oxygen, various non-infectious diseases and immune system activities are inherent sources of stress which need to be controlled.

An understanding of Total Nutrition and of the role of nutricine and nutrient components of feed will extend our knowledge of diet and nutrition and allow us to further develop the connection between health, nutrition and environmental protection. The establishment and maintenance of optimum health and resistance to disease is a function of both nutrient and nutricine intake. Adequate supplies of nutrients and nutricines will counteract the negative effects of the environment and will lead to disease avoidance and good growth and development of animals raised for food.

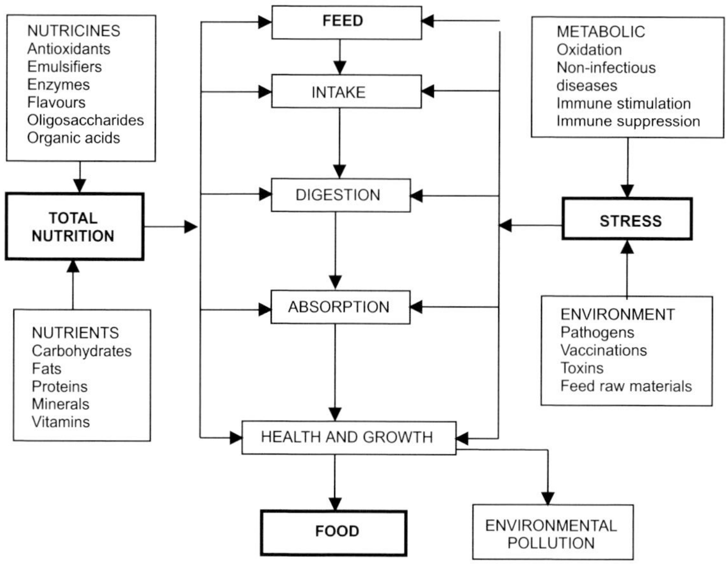

Figure 1
Relationship between Total Nutrition, stress and food production

Total Nutrition is the strategic use of a wide variety of disease avoidance and health maintenance measures which will contribute to an improved and more acceptable system of animal production, without the use of antibiotics. The control of stress, of infectious diseases and of non-infectious diseases involves numerous strategies of which the use of antibiotics is only one, and furthermore antibiotics do not solve problems of viruses and of non-infectious diseases and these syndromes are of increasing importance in animal production. Use of antibiotics will in the future be strictly applied to the treatment of sick animals. The use of antibiotics will then be considered as a final resort in disease control and will only be used when all other measures have failed and not as a replacement of them (Wierup, 2000). Total Nutrition will assist in the maintenance of large-scale, low cost production of food of animal origin which ultimately benefits the human consumer.

Future directions

Nutritional regulation of gene expression

It has long been recognised that nutrition can profoundly alter the phenotypic expression of a given genotype. Indeed a major challenge for modern nutrition is to develop appropriate nutritional programmes to match the improvements in genetics. Insufficient nutrition at a very young age such as in weaned piglets impairs both immediate and long term development. This leads into the concept that nutrition might influence phenotype by regulation of gene expression. This is extremely significant as it suggests another important function for nutrients and nutricines in addition to maintaining health and supporting growth.

A possible mechanism whereby nutrition could influence gene expression is through affects upon hormones and their receptors (Dauncey *et al.*, 2001). Nutrition influences the synthesis and metabolism of many hormones involved in development growth and metabolism. These effects are exerted by specific nutrients, energy status and by changes in feed intake. Many hormones such as growth hormone or insulin interact with receptor molecules to influence the activity of various genes in growth and development. Nutrition can modulate hormone action by regulation of these hormone receptors. Through this interaction with hormones and hormone receptors nutrition plays a key role in the regulation of the numerous genes involved in development, growth and metabolism of animals.

Specific feed components such as conjugated linoleic acid may have a direct effect upon gene transcription which may influence lipid metabolism in the liver and the immune status (Roche *et al.*, 2001). Further research is needed to be able to understand the relative contribution of nutrition and genotype to optimal development. Studies must not only focus upon specific nutrients but also on energy status and overall feed intake. Detailed understanding of nutrition-gene interactions will be necessary for improvements in health maintenance and disease avoidance of animals. Advances in this area will certainly support the concept of Total Nutrition.

Selection for health and disease resistance

Genetics has played a major role in improving animal productivity over the last 50 years and there have been very remarkable and significant improvements in the ability of animals to convert feed into human food items and this can be expressed as a feed sparing effect (Chapter 9). For example in 1996 it was possible to produce 1 kg of eggs with about 500 g less feed than was required in 1972 (Flock, 1998).

Broiler production shows similar significant improvements in efficiency as over the years 1989-96 the FCR of broilers dropped from 1.984 - 1.843 whilst body weights increased over the same period from 1.80 – 2.25 kg (Cobb, 1999).

Clearly animal breeding programmes and genetic improvement have been tremendously successful in production of food of animal origin. Genetics in the future however will need to focus more on strengthening animal health and resistance to disease.

Functional foods

There is considerable scientific evidence from human nutrition that diet can have beneficial physiological and psychological effects that go beyond adequate nutritional effects. There is now great interest in the development and validation of "functional foods" and of the role that diet can play in health maintenance and disease avoidance in the human population. This will require much further research in human nutrition and food science. A particular challenge here is that functional foods are targeted towards healthy persons and therefore evidence of their efficacy will have to be obtained from parameters of biomarkers that underline good health (Roberfroid, 1999).

Total Nutrition will be increasingly used in production of new functional foods for humans such as "designer eggs." The nutritive content of eggs can be significantly improved through supplementing the diet of laying hens with natural feed ingredients. Three egg components have been modified, docosahexanoic acid (DHA), vitamin E and carotenoids. Docosahexanoic acid is an essential omega-3 polyunsaturated fatty acid. It is the most important fatty acid for the development of the brain and retina and for the maintenance of immunocompetence. Vitamin E and carotenoids help prevent the oxidation of the DHA in the egg and in the cell membranes of the body after consumption of the eggs. Vitamin E and carotenoids also protect against heart disease and cancers. Eggs may also easily be enriched in lutein, a natural carotenoid, which, possesses high antioxidant activity and protects the retina of the elderly against loss of vision due to the disorder known as age-related macular degeneration (Olmedilla, 2001).

References

Adams, C. A. (1999). *Nutricines. Food Components in Health and Nutrition*. Nottingham University Press, UK.

Adams, C. A. (2001). Health promoting additives (nutricines). In: *Advances in Nutritional Technology*. Eds: A. F. B. van der Poel, J. L. Váhl and R. P. Kwakkel. Wageningen Pers, Wageningen, The Netherlands. pp. 207-227.

Anderson, E. S. (1965). Drug resistance and its transfer in *Salmonella typhimurium*. *Nature*, **206**, 579-583.

Avery, D. (1999). The fallacy of the organic utopia. In: *Fearing Food*. Eds: J. Morris and R. Bate, pp. 3-18. Butterworth-Heineman, Oxford, UK.

Baker, D. H. and R. W. Johnson (1999). Disease stress, cytokines and amino acid needs of pigs. *Pig News and Information*. **20**, 123N-124N.

Barton, M. D. (2000). Antibiotic use in animal feed and its impact on human health. *Nutrition Research Reviews*, **13**, 279-299.

Bower, C. K. and Daeschel, M. A. (1999). Resistance responses of microorganisms in food environments. *International Journal of Food Microbiology*, **50**, 33-44.

Brandt , K. and Molgaard, J. P. (2001). Organic agriculture: does it enhance or reduce the nutritional value of plant foods? *Journal of the Science of Food and Agriculture*, **81**, 924-931.

Commission of the European Communities (2000). White paper on food safety.

COBB, 2000. Commercial literature.

Dauncey, M. J., White, P., Burton, K. A. and Katsumata, M. (2001). Nutrition-hormone receptor-gene interactions: implications for development and disease. *Proceedings of the Nutrition Society*, **60**, 63-72.

Flock, D. K. (1998). Genetic-economic aspects of feed efficiency in laying hens. *World's Poultry Science Journal*, **54**, 225-239.

Fraser, D. (2001). The "new perception" of animal agriculture: Legless cows, featherless chickens, and a need for genuine analysis. *Journal of Animal Science*, **79**, 643-641.

Jackson, G. J. (1990). Public health and research perspectives on the microbial contamination of foods. *Journal of Animal Science*, **68**, 884-891.

Klasing, K., C., and Johnstone, B. J. (1991). Monokines in growth and development. *Poultry Science*, **70**, 1781-1789.

Knudsen, K. E. B. (2001). Development of antibiotic resistance and options to replace antimicrobials in animal diets. *Proceedings of the Nutrition Society*, **60**, 291-299.

Olmedilla, B., Granado, F., Blanco, I., Vaquero, M. and Cajigal, C. (2001). Lutein in patients with cataracts and age-related macular degeneration: a long-term supplementation study. *Journal of the Science of Food and Agriculture*, **81**, 904-909.

Roberfroid, M. B. (1999). What is beneficial for health? The concept of functional food. *Food and Chemical Toxicology*, **37**, 1039-1041.

Roche, H. M., Noone, E., Nugent, A. and Gibney, M. J. (2001). Conjugated linoleic acid: a novel therapeutic nutrient? *Nutrition Research Reviews*, **14**, 173-187.

Rosen, G., D. (1995). Antibacterials in Poultry and Pig Nutrition. In: *Biochemistry in Animal Feeds and Animal Feeding*. Eds. R. J. Wallace and A. Chesson. pp. 143-173. VCH, Weinheim, Germany.

Saucier, L. (1999). Meat safety: challenges for the future. *Pig News and Information*, **20**, 77N-80N.

Schneeman, B. O. (2000). Linking agricultural production and human nutrition. *Journal of the Science of Food and Agriculture*, **81**, 3-9.

Schumacher, K. D. (2000). Compound feed demand expected to rise to 700 million tonnes. *Feed Tech*, **4**, 40–43.

Thomas, P. C. (2001). Meeting regulatory requirements and consumer demands. In: *Recent Advances in Animal Nutrition 2001*. Eds. P. C. Garnsworthy and J. Wiseman. Nottingham University Press, UK, pp. 1-23.

Vanbelle, M. (2000). Top priority on food safety in the E. U. *Proceedings of the 6th International Feed Production Conference*, Piacenza. Italy. Annex 6.

Wierup, M. (2000) The control of microbial diseases in animals: alternatives to the use of antibiotics. *International Journal of Antimicrobial Agents*, **14**, 315-319.

Wegener, H. C., Aarestrup, F. M., Jensen, L. B., Hammerum, A. M. and Bager, F. (1999). Use of antimicrobial growth promoters in food animals and *Enterococcus faecium* resistance to therapeutic antimicrobial drugs in Europe. *Emerging Infectious Diseases*, **5**, 329-335.

Williams, N. H., Stahly, T. S. and Zimmermann, D. R. (1997a). Effect of chronic immune system activation on the rate, efficiency, and composition of growth and lysine needs of pigs fed from 6 to 27 kg. *Journal of Animal Science*, **75**, 2463–2471.

Williams, N. H., Stahly, T. S. and Zimmermann, D. R. (1997b). Effect of level of chronic immune system activation on the growth and dietary lysine needs of pigs fed from 6 to 112 kg. *Journal of Animal Science*, **75**, 2481–2496.

Witte, W., Jorsal, S. E., Roth, F. X., Kirchgessner, M., Göransson, L., Lange, S. and Pedersen, K. B. (2000). Future strategies with regard to the use of feed without antibiotics in pig production. *Pig News and Information*, **21**, 27N–32N.

2 Virtues of Cleanliness: Feed Quality and Hygiene

Emphasis today is increasingly focussed on "quality" in all activities. The establishment of high standards of food safety and quality is a key policy priority for the European Union and indeed for many other countries. This is epitomised by the creation of the Food Standards Agency in the UK and the proposed creation of a European Food Authority (EFA) by the EU. The EFA is envisaged as an independent body with particular responsibilities for both risk assessment and communication on food safety issues. The term "Food Safety" now also includes animal feed as this is an inevitable link in the whole food quality chain.

Quality of animal feed is a very broad characteristic with several different facets. Feed must clearly have good nutritional quality to support the growth of the animals for which the feed is intended. Feed is the major expense for most intensive animal production systems and consequently feed must allow the animals to achieve the expected performance criteria.

Manufactured feed obviously must conform in its gross nutritional characteristics to the official description of the feed. This is rarely a problem with modern feed formulation techniques and feeds can be routinely produced in large volumes with the required nutritional quality for the target animal species. In general modern feed mills should be able to dose feed ingredients within 1% of the target weight and with a variation of less than 3% (van Kempen et al., 2001).

Feed quality also can have an impact upon the health and welfare status of the animals and this becomes increasingly important in modern times. Feed quality must also ensure that the human food products obtained from animals are safe and of high nutritional value. The safety of food from animal origin has become paramount after various food safety scares from Salmonella, BSE and dioxins.

Feeds must have many specific characteristics in order to be perceived generally as of good quality. They must be free from any significant contamination by a whole host of various undesirable substances (Table 1). These have a wide range of chemical composition and diverse origins. They include; moulds and secondary products of moulds, the mycotoxins, bacterial contamination, seeds and toxic compounds in plants, environmental chemical pollutants such as heavy metals, dioxins and pesticides. The fat in feeds must be stable and not oxidised. Pellets must be sufficiently durable to survive transport and automatic feeding systems. Some feeds are required to have an extended shelf life without suffering deterioration. The use of natural versus synthetic raw materials is increasingly desired.

Given all these demands and potential threats to feed quality perhaps it is surprising that so much feed is produced to such generally high standards. Nevertheless consumer attitudes and legislative pressures continue to insist on even higher standards and ever-lower risks.

Micro-organisms and secondary metabolites	Plant seeds and toxic compounds	Environmental pollutants
Moulds:	Seeds:	Dioxins
Aspergillus spp.	Castor-oil	Fluorine
Fusarium spp.	*Crotalaria* varieties	Heavy metals
Penicillium spp.	*Lolium remotum*	Arsenic
Mycotoxins:	*Lolium temulentum*	Cadmium
Aflatoxins,	Toxic compounds:	Lead
Ergot	Alkaloids	Mercury
Fumonisins	Biogenic amines	Pesticides:
Ochratoxin,	Glucosinolates	Aldrin
Tricothecenes	Gossypol	Dieldrin
Zearalenone	Isothiocyanate	DDT
Bacteria	Theobromine	Endrin
Salmonella spp	Trypsin inhibitors	Heptachlor
Clostridia spp		

Table 1
Some undesirable substances that may occur in animal feeds

Feed consumption however is another paradox of life. Clearly feed is essential to support life and growth of animals. However feed is also a major route for the delivery of toxins and of pathogenic micro-organisms to an animal and it is extremely important that feed raw materials and manufactured feeds are of high quality in terms of nutrition and in terms of microbial contamination. The conviction that safe feed is a major contributor to safe food is now accepted by the EU and will dictate many aspects of animal nutrition in the future. Consequently feed manufacturers, farmers and food processors all have a major responsibility for food safety. For animal production and feed manufacture this implies an enormous range of demands and conditions. In addition the general reluctance to use antibiotics and other pharmaceutical products in animal production increases the significance of feed quality and hygiene. Total Nutrition recognises that to raise animals with minimal reliance upon drugs and medicines high quality animal feed must be produced and due care and attention must be paid to manufacture, storage and transport of both feed raw materials and animal feeds.

Destruction of feed quality

To satisfy this multitude of demands and requirements, is no easy task for the nutritionist and feed manufacturer. Feed raw materials and feeds are materials of natural origin and are inevitably sensitive to various destructive chemical and biological processes as illustrated in Figure 1. The major destructive chemical process is oxidation. Many different uncontrolled oxidation reactions, usually termed autoxidation, occur in raw materials and in feeds which results in the destruction of important molecules and in the generation of rancidity and oxidative stress. This in turn will have an impact upon feed intake and animal health and performance.

Autoxidation of feed lipid components is a major cause of reduction in feed quality, affecting nutritive value, taste, aroma, colour and texture. Consequently autoxidation must be avoided during the conservation of feed raw materials and in the manufacture of animal feeds. The problems and dangers of autoxidation of feeds are confounded also with problems of oxidative stress and this is more fully discussed in Chapter 7.

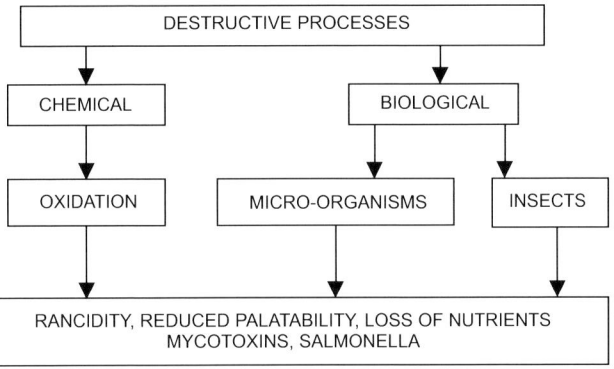

Figure 1
Problems of raw material and feed quality

The major concerns in feed quality are to avoid any toxic ingredients, to avoid oxidative rancidity, to avoid contamination by moulds and bacteria and infestation by insects. These problems can lead to substantial losses of dry matter and of nutritional value due to the activities of insects and moulds and this may lead to further problems of mycotoxin production. The activity of insects can in the worst cases absolutely destroy feed raw materials.

In most raw materials or manufactured feeds bacterial activity will not be very significant as these materials are too dry to support growth of bacteria. However pathogenic bacteria can survive in feeds and this becomes a source of infection for the animals consuming the feed. In many countries, including all member states of the EU, contamination of feeds by pathogenic micro-organisms such as *Salmonella* or *Clostridia* is also subject to legal controls.

Autoxidation in feeds

Autoxidation and its control are important factors in the production and conservation of feed raw materials and in manufactured animal feeds. Oxidative stress and its control is also important to protect animal health and to improve animal performance (Chapter 7). Therefore we must consider oxidation at two levels; the direct effects upon feeds and the physiological effect upon animals.

There are many lipid or fat components of feeds which spontaneously react with atmospheric oxygen and suffer deterioration in the process of autoxidation. These include; fats and oils, mono- and diglycerides, and sterols. Vitamins A, D, E and K are fat-soluble and are also sensitive to autoxidation. Carotenoid pigments, lutein, zeaxanthin and astaxanthin are important in colouring poultry and fish and carotene carotenoids are important lipid components of silages and hay and give colour to dairy products. Essential oils such as orange, peppermint or anise are used in many flavours that are incorporated into both human foods and animal feeds. Emulsifiers, which may be various lecithins and lysolecithins are used in many manufactured foods and are often incorporated into piglet and fish feeds and into calf milk replacers.

Fats and oils together with other lipid materials are important and also relatively expensive components of animal feeds. Fats and oils are the greatest contributors of energy per unit of weight of any feed ingredient. Consequently it is economically important that expensive feed ingredients are not destroyed by autoxidation.

Autoxidation of feed lipid components generates an oxidative rancidity that is a major cause of reduction in feed quality, affecting nutritive value, taste, aroma, colour and texture. These in turn will influence animal performance and some by-products of oxidative rancidity are detrimental to health.

The autoxidation processes are fairly complex but have been reasonably well established and can be described as a free radical chain reaction. (Adams, 1999; Duthie, 1993). There are three steps in the chain reaction which may be illustrated as follows:

1) **Initiation:**   RH ⟶ R·

This produces a free radical (R·) and may be promoted by the presence of metals, light and temperature

2) **Propagation:**   R· + O_2 ⟶ ROO·

ROO· + RH ⟶ ROOH + R·

During this step oxygen is consumed, there is a continuous propagation of free radicals and production of peroxides (ROO·) and hydroperoxides (ROOH).

3) **Termination:**   ROO· + ROO· ⟶ Stable end products

The stable end products are the chemical manifestation of rancidity and are a mixture of aldehydes, ketones and low molecular weight fatty acids. They tend to have unpalatable tastes and unpleasant odours.

Knowledge of the chemistry of autoxidation has allowed the development of antioxidants to control the destructive oxidative processes. Indeed development of antioxidants to control autoxidation is a major preoccupation for the feed, food, plastics and rubber industries, all of whom deal with products susceptible to autoxidation.

Autoxidation of feeds and the development of oxidative rancidity may be avoided or retarded by the use of various proprietary antioxidant products available for use in fats and oils and to stabilise vitamin premixes. These can be of synthetic origin such as BHA (butyl hydroxy anisole) or BHT (butyl hydroxy toluene) or of natural origin such as rosemary extracts, or tocopherols.

Consumption of oxidised feeds

Oxidised feeds are likely to be poorly consumed by animals leading to poor growth rates and they are also a potential health risk. Oxidised feeds will have poorer organoleptic characteristics than fresh feeds but also may have lost nutritional value if fats and vitamins have been destroyed. Both of these factors will contribute to the reduced growth and performance of animals consuming oxidised feed that has been reported for several species.

Rats showed significantly lower feed intake and body weight gain on an oxidised diet compared to a diet which was not oxidised (Lamghari *et al.*, 1997). Rats were fed a lupin-based feed that was stored for 15 days and then heated which oxidised the unsaturated fatty acids in the feed. Similar reductions in feed consumption and in growth rate were seen in rats fed thermally oxidised maize oil (Nwanguma *et al.*, 1999).

Inclusion of oxidised vegetable oils into broiler feeds had quite marked influence on growth rates (Engberg *et al.*, 1996). This was evident at 24 days of age and continued until 38 days (Table 2). At 38 days of age broilers fed oxidised oil had an average body weight 109g or some 5% lower than broilers fed fresh oil. The retention of fat, energy and a-tocopherol was lower in birds fed oxidized fat. This indicates that feeding oxidized fats has a variety of different effects upon metabolism.

Table 3
Influence of oxidised vegetable oils on body weight (g) of broiler chickens

Age	Treatment	
(days)	Fresh oil	Oxidised oil
24	1015	950
31	1573	1466
38	2092	1983

Note: fresh oil, peroxide value 1 meq./kg oil, oxidised oil, 156 meq./kg oil.

Dietary fats, particularly fish oil, is an important feed ingredient for mink and other fur bearing animals. Fish oil is however very sensitive to oxidation and oxidised fish oil in the diet reduces digestibility of several nutrients including gross energy and crude fat (Borsting *et al.*, 1994). The performance and health of the animals was also seriously affected by oxidised oil in the diet (Engberg and Borsting, 1994). Animals fed oxidised fish oils lost weight and those fed the highest levels of oxidised fish oil were extremely thin. The weights of the liver and heart, as a proportion of body weight, were increased by feeding oxidised fish oils.

Nutritional studies with different animal species demonstrate that feeding oxidised fats and oils results in reduced feed consumption and has undesirable physiological effects. Consequently the protection of dietary fats and oils against autoxidation is very important for production of good quality feeds and this can be done by the judicious use of various antioxidants.

Antioxidants are valuable nutricines but they have to be used correctly. Ideally feed raw materials susceptible to oxidative destruction should be treated with appropriate levels of antioxidants as early as possible in the manufacturing process. Antioxidants will not improve the flavour of feeds already oxidised, neither can they reverse the processes of autoxidation. Antioxidants prevent, not cure autoxidation. They will not prevent the formation of free fatty acids which arise from chemical hydrolysis of fats.

In Total Nutrition protection of lipid materials against the destructive processes of autoxidation is of fundamental importance in maintaining feed quality.

Growth of moulds and insects

Raw materials and feeds are inevitably contaminated by a wide range of moulds and may frequently be infested with insects. Furthermore there is a complex relationship between mould and insect contamination. In general grain samples heavily infected with storage moulds tend to have significant insect populations. The presence of insects may also predispose stored grain to rapid infection by moulds, either by damaging the grain and making it more vulnerable to attack by moulds, or by acting as carriers of the moulds.

Protection of raw materials and feeds against mould contamination is difficult because many feed raw materials such as cereals and oilseeds are harvested seasonably and therefore have to be stored and often transported before use from sites remote from feed manufacture. They will commonly be stored for several months and may be stored for several years. They frequently need to be transported thousands of kilometres. The storage and transport conditions of raw materials are important as there is always the danger of contamination by moulds.

In order to avoid and to control mould and insect contamination the environmental conditions necessary for survival and growth of these organisms must be understood. Strategies can then be devised to ensure the safety and to maintain the nutritional value of animal feeds.

Oxygen

Both moulds and insects are aerobic organisms and most raw materials and feeds are stored under normal atmospheric conditions where oxygen is present. Some feeds such as silages for example are deliberately stored under anaerobic conditions but these are still subject to aerobic spoilage when the silage clamp is opened. In general it is not possible to control mould growth in feed raw materials or in feed by excluding oxygen.

Temperature

Both moulds and insects will grow more rapidly as the temperature increases. Given the large volumes of raw materials needed for feed manufacture it is practically impossible to store these materials at low temperatures in order to avoid spoilage by micro-organisms and insects. Certainly in more northern regions with cool winters growth of moulds and insects will be less of a problem than in warmer or tropical areas.

Time

Growth of moulds and insects requires time but this may be a matter of days or weeks under suitable conditions. On the other hand dry raw materials stored under good conditions may retain good nutritional quality for many months or even years.

Moisture

For the conservation of raw materials and successful storage of feeds, moisture content is the most important factor. In practice the moisture content of a mass of grain or other raw material is determined from representative samples and it is assumed that this average value is uniform throughout the grain mass. Often it is also assumed that this average moisture content will remain stable, but this is unlikely to be true in practice. The moisture content of the individual cereal grains will vary, the moisture content of different lots of grain will inevitably be different. In reality a bulk mass of feed grains is a mosaic of different moisture contents and is an inherently unstable system.

For safe storage of feed raw materials the highest moisture content that prevails in any portion of the bulk mass is important, not the average moisture content. The highest moisture content will determine how fast grain will deteriorate in storage. Furthermore relatively small changes in moisture content or temperature can rapidly influence the growth of moulds or insects.

This moisture is also easily released from the grain by changes in temperature. Consequently during storage or transport there will be an inevitable and constant movement of moisture from one region of a grain mass to another.

Moisture content is an important factor in the activity of both moulds and insects. Rice and maize weevils for example cannot reproduce in grain with a moisture content below 9%. Flour beetles on the other hand can produce progeny in flours or in grain dust that are extremely dry.

During larval development and growth, an insect produces metabolic water and heat. This makes the environment in the stored raw material more hospitable for mould growth. The interaction of moisture content, moulds and insects can rapidly lead to spoilage of stored feed raw materials. Bulk masses of grain or feeds with moisture contents as low as 11% may develop symptoms of heating (Cotton and Wilbur, 1982). This heating of relatively dry materials is most likely caused by the metabolism of insects and can result in temperature increase up to 42°C.

If heating occurs in stored materials with a moisture content of 15% or more this is likely due to the growth of moulds and can produce temperatures as high as 65°C.

Mould growth can readily develop in stored materials if the moisture content in any one area rises above 13% and so insect activity is usually very beneficial to mould growth. The increase in moisture content and temperature due to growth of insects is frequently followed by rapid growth of moulds.

In materials with 11-15% moisture a combination of the two sources of heating probably occurs. A small local rise of temperature in a "hot spot" will accelerate the metabolism of the insects and speed up the rate of population increase. The temperature and moisture increases resulting from insect metabolism will then create an environment very favourable for mould growth.

Insects may also be carriers of moulds in their intestinal flora (Table 4). *Aspergillus, Penicillium*, and *Cladosporium* species are common moulds contaminating stored grains and other feed raw materials (Fleurat-Lessard, 1989). Therefore insect infestation may well help to distribute moulds widely throughout a bulk mass of grain or feed.

Table 4
Mould species contaminating the intestinal tract of insects

Insect species	Mould species
Sitophilus oryzae *S. granarius* (weevil)	*Aspergillus flavus* *A. candidus*
Oryzaephilus surinamensis (grain beetle)	*A. flavus, A. niger, A. ochraceus* *A. versicolor, Penicillium rugulosum* *Cladosporium* species
Tribolium castaneum (flour beetle)	*A. flavus, A. niger, A. ruber* *A. sydowi, A. glaucus, A. ochraceus* *P. islandicum*

Moisture migration, brought about by temperature effects, will stimulate mould and insect growth. Insect growth will further stimulate mould growth. Mould growth will generate heat and moisture accelerating the growth of more moulds leading to a widespread contamination by moulds and insects with an increase in temperature and moisture content of the stored materials.

Insects and moulds utilise stored raw materials as a food source for their own growth and so will reduce the nutritive value of stored grains. The presence of insects and mould will give feed raw materials unpleasant odours and flavours so reducing palatability or acceptability of these materials. Raw materials heavily infested by insects or contaminated by moulds will tend to form clumps. There may be the formation of bridging in silos and a general difficulty to move materials by automatic transport systems.

Water activity (a$_w$) and feed moisture content

Moisture level in manufactured feeds, has long been an important quality parameter and in many countries legislation limits the moisture content of feeds. However in modern animal nutrition it is increasingly important to consider moisture content of feeds from several points of view. Water in feed influences microbial activity and insect infestation, palatability and intake, colour, texture and mechanical properties. These in turn have an impact upon nutrition and digestibility of feeds, pellet quality, shelf-life of feeds and economics of feed manufacture.

Moisture content of raw materials and of feeds is traditionally expressed as percentage by weight. This moisture content is in reality the amount of water that can be driven off by heating the feed in an oven usually at 105°C. However gross moisture content is not the controlling factor in feed conservation. The true parameter is the actual amount of water in a material that is available to support growth of micro-organisms. If the water present is not available, then growth of micro-organisms cannot occur. This is commonly seen in molasses where growth of micro-organisms does not occur despite the high moisture content unless the molasses is diluted with additional water.

For effective control of moisture in feed it is necessary to consider both the quantity and the quality of water in a feed. Gross moisture content of a feed is a quantitative measure but this does not give much information about the qualitative nature of the water in the feed and its availability for chemical activity, as a nutrient, or to support growth of micro-organisms, especially moulds. To understand more completely the interaction of water and the growth of micro-organisms both the gross moisture content and the water activity (a$_w$) must be considered.

Water activity can be defined as the ratio of the vapour pressure of water in a feed (P$_{feed}$) to that of pure water (P$_0$):

Water activity (a$_w$) = P$_{feed}$ / P$_0$

Pure water has an activity value of 1.00 and the scale runs from 0-1.00. Osmotic and other attractive forces in a complex mixture such as a feed usually lower the a_w below that of 1.00. Water activity is also related to the relative humidity of air above a sample in a sealed container:

Water activity (a_w) = relative humidity(%) / 100

This means that if a feed sample is sealed in an airtight container the humidity of the air in the headspace above the feed will rise to a stable or equilibrium value of perhaps 67% which means that the a_w of the feed was 0.67. Essentially a_w is a measure of the degree to which water is bound within a feed and will not be available for further chemical or microbiological activity.

Micro-organisms require the presence of available water for growth and metabolism, and this available water is best measured by a_w. Micro-organisms differ in their response to a_w, generally yeasts and moulds can grow at a lower a_w than bacteria. Most bacteria require a_w values above 0.85 for active growth, but some moulds and yeast are capable of growth at a_w values as low as 0.60. At a_w values below 0.55 DNA structures are disrupted and living cells can no longer survive (Enigl and Sorrells, 1997).

Water activity has long been used in the food and pharmaceutical industries to indicate the amount of available water in a product and to develop programmes and products to control the growth of spoilage micro-organisms. Altering a_w in food is usually done by adding various water-soluble materials such as glucose, sucrose, syrups or salt. These can all reduce the a_w as illustrated in Table 5 and are valuable conservation agents for human foods.

Table 5
Variation in water
activity (a_w) with
various solutes

Water activity	Solute concentration (%w/w)			
	Sodium chloride	Glucose	Sucrose	Glucose syrup
1.000	0	0	0	0
0.990	1.74	8.90	15.45	3.15
0.900	14.18	48.54	58.45	31.49
0.860	18.18	58.45	65.63	44.08
0.753	26.50	-	-	-

Animal feeds are not usually conserved with these materials and consequently a_w is relatively little used in animal feed quality control. However the a_w is of value in terms of moisture control of feed as it can ensure that feeds are produced to the maximum moisture content without risk of microbial spoilage. This has considerable economic importance as moisture losses during raw material storage and feed manufacture are responsible for significant shrinkage of production volumes, sometimes up to 3%. This production loss must either be borne by the producer as an additional operating expense or must be avoided by efficient control of moisture content in the finished feed.

Mycotoxins

All living organisms including moulds produce a large number of different molecules during their growth which are referred to as metabolites. In moulds metabolites can be conveniently divided into primary and secondary metabolites. Primary metabolites are essential for the life of the organism and include compounds such as sugars, amino acids and lipids. Secondary metabolites are substances not essential for life and growth of the organism and include the mycotoxins. There is no direct association between microbial growth and production of secondary metabolites. Frequently however these secondary metabolites, such as mycotoxins, have an adverse effect on other organisms.

Since mycotoxin-producing moulds grow on staple raw materials for animal feeds and human foods, both the animal and human populations are affected by them. Therefore mycotoxins must be considered persistent and common contaminants of feed and foods. The occurrence of mycotoxins in feeds and human foods is an animal and public health problem of major concern worldwide. Extreme vigilance and careful management is vital to ensure that levels of mycotoxins are kept acceptably low.

Mycotoxins cause a whole of array different adverse effects on health and production of animals depending on the quantities of mycotoxin that an animal receives:

1) Acute primary mycotoxicosis: Occurs with high levels of mycotoxins. Animals show specific overt signs of disease and of increased mortalities.

2) Chronic primary mycotoxicosis: Occurs with moderate or low levels of mycotoxins. Animals show reduced growth and performance without overt signs of disease or of increased mortalities.

3) Secondary mycotoxicosis: Occurs when small amounts of mycotoxins are consumed, particularly mixtures of different mycotoxins. Animals are predisposed to infection and disease through suppression of their immune system.

The complex processes involved in mycotoxin production by moulds make it difficult to predict which toxin will be produced, when it will be produced and in what concentration. Unfortunately it is not possible to prevent entirely the production of mycotoxins before the harvest of agricultural crops, in storage, or during processing operations.

Cereal grains can be initially contaminated by so-called "field moulds" such as, *Alternaria, Cladosporium, Fusarium* and *Penicillium*, whilst developing on the plants in the field. Field moulds, except for *Penicillium*, require high relative humidity and water content and are not able to compete in storage conditions. They become dominated by species of "storage moulds" such as *Aspergillus, Monascus, Mucor, Penicillium,* and *Wallemia.* These various mould species can collectively produce several hundred toxic metabolites, and the major contaminants of feed raw materials are listed in Table 6.

Mycotoxin	Mould species
Aflatoxins,	*Aspergillus* spp.
Fumonisins	*Fusarium moniliforme*
Ochratoxin,	*Aspergillus* and *Penicillium* spp.
Tricothecenes	Several mould spp.
Zearalenone	*Fusarium* spp.

Table 6
Major mycotoxins found in feed raw materials and associated toxigenic mould species

Mycotoxins may not necessarily occur in feeds one at a time. It is more likely that mould contamination of feeds could lead to the production of several mycotoxins. A detailed study of maize products for animal feeds showed that many samples contained a mixture of multiple toxins including trichothecenes, fumonisins and moniliformin (Scudamore *et al.,* 1998). It is probably the presence of several different mycotoxins at low levels in feeds which causes animal health and performance problems. In experimental trials a combination of aflatoxin and moniliformin very severely reduced body weight gain in chicks over 21 days (Table 7), Kubena *et al.,* 1997).

Treatment		Performance characteristics	
Aflatoxin (mg/kg)	Moniliformin (mg/kg)	Weight gain (g)	FCR
0	0	736a	1.60b
3.5	0	638b	1.65b
0	100	515c	2.10a
3.5	100	490c	1.91a

Table 7
Effect of feeding diets containing moniliformin and aflatoxin on performance of broiler chicks over 21 days

An insidious effect of many mycotoxins is that they cause immunosuppression in animals (Li *et al.,* 2000; Weidong, 1991) . This will make them more susceptible to various infectious micro-organisms.

This has been well demonstrated in poultry when ochratoxin A was given to broiler chicks together with *Salmonella* (Elissalde *et al.*, 1994). At 21 days there was quite significant mortality in those chicks infected with *S. typhimurium* and also given ochratoxin A in the feed. Chicks at this age would normally not be sensitive to *Salmonella* but the presence of the mycotoxin probably suppresses the immune system of the broiler chicks and they become more sensitive to pathogenic bacteria.

Mycotoxins are generally quite stable molecules and may remain in raw materials and in human food products for a considerable time. Mycotoxins may be found in feed raw materials long after the moulds that produced them have died. This has several serious consequences for human and animal health. When an animal consumes a mycotoxin in the feed such as aflatoxin, it will be absorbed from the gastrointestinal tract and distributed by the bloodstream throughout the body. Studies on the distribution of aflatoxin in chickens revealed that 6.2% was retained in breast and leg meats (Mabee and Chipley, 1973). Mycotoxins are usually resistant to normal conditions of feed manufacture and to cooking processes since they are relatively heat stable as shown in Table 8 (Egon Josefsson and Möller, 1980; El-Banna and Scott, 1984). Clearly for safe food production mycotoxin contamination must be minimised.

Food product	Cooking temperature (°C)	Cooking time (min)	Ochratoxin A remaining (%)
Faba beans	115	120	80.8
Wheat	100	30	94.1
Pig kidney	160	5	76.0
Pig fat	150	12	100.0

Table 8
Amount of ochratoxin A remaining in food products after heat processing

Prevention of mycotoxin contamination

Mycotoxins only occur if there has first been mould contamination of raw materials or feeds. The first line of defence is clearly to reduce mould contamination and growth to the minimum. The judicious use of organic acid based mould inhibitors can prevent excessive mould contamination and even reduce existing mould levels. This can be very important in raw materials that are stored for some time. The most effective products need to be based on acids such as propionic and sorbic. It is curious that formic acid is not as inhibitory to mould growth as is propionic acid and may allow aflatoxin production in stored materials. This has been demonstrated experimentally by the treatment of barley with propionic and formic acids after inoculation with *Aspergillus parasiticus*. Treatment with formic acid promoted the development of aflatoxin above the control, whereas treatment with propionic acid completely inhibited mycotoxin production (Holmberg *et al.,* 1989).

The relatively poor mould inhibiting effect of formic acid has led to a change in EU legislation where formic acid (E236) or mixtures of acids with more than 50% formic acid are prohibited for use as grain preservatives.

Decontamination

Certain environmental conditions close to harvest time may result in unavoidable contamination of cereals with mycotoxins. Consequently there has been considerable research effort devoted to possible decontamination strategies for mycotoxins. The general characteristics of mycotoxins such as their occurrence in small amounts (frequently <1 g/tonne), their great chemical diversity and stability are formidable obstacles for the development of decontamination procedures. It is unlikely that any one decontamination procedure would be effective for all the different mycotoxins that may contaminate raw materials. Decontamination procedures also need to be reasonably cheap and capable of treating large volumes of raw materials. They should not alter the desirable nutritional characteristics of the raw materials.

Various physical methods such as milling and heating have not found great practical application. Cooking or other heat processing systems used in animal feed production are not adequate to destroy all mycotoxins since they are relatively heat stable as illustrated in Table 8. Even high temperatures as used in food cooking processes only destroyed a small amount of the ochratoxin A. Animal feed processing usually has much lower temperatures and so it is unlikely that the feed manufacturing processes would have much effect upon mycotoxin content of contaminated raw materials.

Numerous chemicals have been tested for their ability to decontaminate raw materials containing mycotoxins. Some examples are sodium bisulphite, dimethylsulphoxide, calcium hydroxide/monomethylamine, sodium hypochlorite, hydrogen peroxide, methanol, ammonium bicarbonate, water, sulphur dioxide, formaldehyde, chlorine gas and anhydrous ammonia.

The trichothecenes are water-soluble and can be removed by soaking and washing contaminated grains. Solutions of sodium bisulphite have also been successfully used to treat cereals contaminated with deoxynivalenol (DON). Treatment of contaminated cereals with formaldehyde solution can reduce zearalenone levels.

The chemical process most widely used is that of ammoniation, which is operated on an industrial scale in specifically designed plants in France, Senegal, India and the UK. This is quite an effective process in reducing aflatoxin contamination of raw materials. Treated materials usually retain a strong odour of ammonia which is unpleasant to humans. However incorporation of such treated raw materials into animal feeds has been quite successful.

Reducing mycotoxin effects in the animal

The ideal scenario is to avoid feeding mycotoxins to animals but nevertheless this is not so easy to control. Feed raw materials and finished feeds readily support mould growth and mould contamination is very widespread in the environment. Very low levels of mycotoxins may be present in feeds which will be difficult to detect by chemical analysis. Consequently in Total Nutrition it will be important to reduce the effects of mycotoxins in the animal by nutritional means wherever possible.

Micro-organisms in the gastrointestinal tract of animals are an important detoxifying system for the animal. This is probably the reason that ruminants are more resistant to the effects of mycotoxins than monogastric animals as the toxins may be metabolized by the rumen micro-organisms. The development of nutritional programmes to promote the activity of micro-organisms able to degrade mycotoxins in the gastrointestinal tract could be a useful research topic.

A second strategy is to incorporate into feeds materials which may absorb the mycotoxins and retain them in the gastrointestinal tract. This would prevent their absorption into the animals' body and nullify the toxic properties of the mycotoxins. This strategy has been the subject of considerable research in recent years. Various absorbing materials have been used including charcoal, zeolites, bentonite, spent bleaching clay from vegetable oil refining, and hydrated sodium calcium aluminosilicate (Bailey *et al.*, 1998; Ramos and Hernandéz, 1997; Schell *et al.*, 1993). The main problem here is that mycotoxins are chemically quite diverse and so no single toxin- binding compound will be effective for all mycotoxins.

Mycotoxins absorbed from the gastrointestinal tract are usually metabolised in the liver and there is already some evidence that components of feed may be beneficial in ameliorating the effects of mycotoxins in the animal body. Protein or the amino acid cysteine supplementation assists the biochemical detoxification of aflatoxin and alleviates some of the adverse effects in the animal. This may be due to the activity of a glutathione (GSH) detoxification system in the liver. GSH is a tripeptide which contains cysteine and it binds aflatoxin in the liver and renders it non-toxic after which it is excreted in the bile, and subsequently passes into the urine.

Ascorbic acid provides considerable protection against the toxic effects of ochratoxin A in laying hens. The mechanism of the effect is not known but it may be related to the ability of ascorbic acid to reduce the production of lipid peroxides. Promotion of lipid peroxidation in tissues is one of the ways in which ochratoxin A produces its toxic effects. Good levels of antioxidant protection in feeds may have additional benefits in helping overcome mycotoxin problems as well as preventing oxidative destruction of nutrients.

Administration of N-acetylcysteine, an acetylated derivative of the amino acid cysteine, was able to prevent the effects of aflatoxin toxicity on weight gain and diminished the severity of histological lesions in broilers (Valdivia *et al.*, 2001). This is quite an interesting finding as N-acetyl cysteine has been widely prescribed to humans in several countries and therefore its safety and pharmacological properties are well established. It is an excellent source of sulfhydryl groups and is capable of stimulating synthesis of glutathione which is also probably involved in reducing the effects of mycotoxins. N-acetyl cysteine may be a very useful compound to control aflatoxicosis in broilers.

Bacterial contamination

Many bacterial species are important pathogens in human foods and have to be strictly controlled. There must be a perennial campaign to avoid contamination of human foods by organisms such as; *Campylobacter, Clostridia, E.coli, Salmonella* and *Yersinnia*. These organisms can occur in foods of animal origin such as, meat, milk and eggs. A major objective of Total Nutrition is the production of safe human food and therefore animal production must be directed towards the avoidance of bacterial pathogens in the final food products.

Animal feed is generally too dry to support the growth of bacteria and so is unlikely to be a major source of food pathogens. *Campylobacter* for example is an important pathogen of fresh chicken meat but is quite sensitive to dry conditions and is not thought to be transmitted to broilers via the feed. However some bacterial species such as *Clostridia* and *Salmonella* survive in dry conditions and feed can be a potential source of contamination by these organisms. Feed has been considered as one of the vectors of salmonellosis in animals and humans. Consequently attention is now strongly focused on control of *Salmonella* in animal feeds and in the animals themselves.

In an effort to reduce contamination many feed mills and raw material production sites have instituted routine monitoring programmes and associated feed hygiene programmes. This may include the implementation of GMP (Good Manufacturing Practice) codes or the more rigorous HAACP system (Hazard Analysis of Critical Control Points).

Continuous efforts must be directed towards reducing the overall load of *Salmonella* in the environment of animals. Reducing the potential of feed and feed ingredients to act as a source of *Salmonella* organisms, which can seed the environment for subsequent recycling by birds, rodents, other animals and insects continues to be a necessary and prudent course of action.

Monitoring individual rations for *Salmonella* however is a costly and an inefficient process because of the need to sample large volumes of material with a heterogeneous distribution of contamination. This is particularly true of cereal raw materials which form a large part of all feeds. Contamination of grains with faeces of rodents, wild birds and cats containing *S. typhimurium* DT104 has been found frequently on cattle and pig farms where grain stores are close to infected livestock housing. However surveys of raw materials and feeds do not give the complete picture and provide little information as to the origin of the contamination.

Therefore assessing contamination of the feed manufacturing process is a more valuable control programme (Davies and Wray, 1997). The microbiological quality at different stages throughout the process should be checked using swabs and samples for analysis by a microbiological laboratory. Key critical points for assessment are:

- Intake pits for raw materials
- Raw material silos
- Pneumatic transfer systems
- Elevator boots
- Hammer mill
- Mixer
- Cyclones
- Pellet press
- Transfer from the pellet press outlet to the cooler inlet
- Cooler
- Finished product silos
- Bagging out area

Samples of spillage should be collected from key critical points in sterile plastic jars or bags. Samples of dust should be taken around intake pits, from cyclones and storage silos. Aggregated material from within coolers, particularly under the cover or lid of coolers should be sampled. Animal droppings can also be collected when present.

The value of such a programme is that it identifies locations in the feed manufacturing system that may be sources of contamination. Areas around the intake pits and around the pellet coolers will be the most likely places where *Salmonella* could reside. Once these locations are identified then rigorous hygiene measures can be implemented to control these sources of contamination.

Salmonella control in raw materials and feeds

The control of *Salmonella* in animal production starts with feed raw materials and feeds. Here the major problem is most likely to be contamination which occurs after production since many processes such as solvent extraction of vegetable oils from oilseeds, heating of protein meals and high temperature pelleting may destroy any *Salmonella* initially present. Contamination after production of the raw materials and feeds from rodents, birds or the environment is likely to be local in distribution and not homogeneously dispersed through the bulk of the material.

Salmonellas are quite robust micro-organisms and consequently they will survive in many feed raw materials and through feed manufacturing processes. The control of *Salmonella* contamination in raw materials and feeds is rendered more difficult due to the ability of *Salmonella* to survive in dry materials and also to the heterogeneous nature of the contamination. This means that once feeds or the environment are contaminated by *Salmonella* there is a danger of this contamination persisting for a long time. For example *S.enteritidis* and other serotypes persisted in poultry feed for at least 10 months (Davies and Wray, 1996). There was also long term persistence of *S. enteritidis* in commercial broiler breeder and layer breeder units. In one broiler breeder unit *S. enteritidis* was found in samples of old fan dust outside a poultry house 20 weeks after it was depopulated. On a layer breeder site *S. enteritidis* was isolated from 36.7% of the environmental samples taken 26 weeks after depopulation.

When *Salmonella* are subjected to storage at low water activity even for relatively short periods the resistance of the organisms to heat is markedly increased (Kirby and Davies, 1990). Consequently in relatively dry materials such as animal feeds low pelleting temperatures are frequently not adequate to consistently kill-off all salmonellas. Pelleting at temperatures of 80-82°C has a beneficial effect upon reducing *Salmonella* contamination. However heating poultry feed to 71°C for 80 second did not kill-off all *Salmonella* (Matlho et al., 1997). Not many raw materials are routinely pelleted or heat treated but where this is done for efficient *Salmonella* control it needs to be consistently above 80°C.

Heat treatment at the appropriate temperatures will clearly destroy *Salmonella* but this is not always a practical possibility. In general layer feeds are not pelleted and whole wheat may be incorporated into broiler feeds without any heat treatment. In these cases *Salmonella* control must be maintained by using chemical treatments. In very sensitive situations such as poultry breeder feeds both heat and chemical treatments may be used.

Products intended for control of bacterial contamination of feeds must be non-toxic at the concentrations used and remain undegraded in the feed until it is consumed. Proprietary products based on short chain fatty acids such as formic, acetic, lactic, propionic and sorbic fulfil these criteria and are widely available. In some member states of the EU and other countries treatment of raw materials is tolerated with products containing formaldehyde together with organic acids and this is a very powerful antibacterial agent. In the USA broiler feed may be treated with formaldehyde against *Salmonella* but this is not permitted in the EU.

These organic acids, with the exception of formic acid are the same ingredients used for mould inhibition. An importance difference however is that much higher dosages are used for *Salmonella* control than for mould control. Moulds at low levels are always present in feeds, and providing they do not produce mycotoxins are not considered particularly dangerous. The objective in a mould inhibition programme is to prevent further mould growth and this can be accomplished by the use of relatively low dosage rates of a mould inhibitor product of 0.5-2.0 kg/tonne. Salmonella control is quite different in that no salmonellas at all are tolerated and the objective here is to eliminate all salmonellas from the treated materials. This requires fairly high dose rates of products, usually from 2-10 kg/tonne depending upon the material treated.

Proprietary bacterial inhibitors are available as liquid and dry products. Both can be used for feed treatment although the liquid products are generally used in raw material treatments. Bacterial inhibitors frequently have formic acid as a component which has a boiling point of 100°C and so is more volatile than propionic acid. Again high temperature processing will cause the loss of some formic acid but usually not all will be lost.

Chemical treatment of raw materials against *Salmonella* contamination is not a simple matter. There are two major problems with raw materials. The worst problem is dust formation. Raw materials may be ground, they certainly must be transported through mechanical handling systems, weighed and distributed. These processes inevitably generate dust. Another important problem is the condensation of water on the inside of windows or walls where it mixes with the dust. This mixture can allow

Salmonella to grow because of the raised water activity. As the *Salmonella* grow at a very wide range of temperatures from about 10°C up to 44°C growth and multiplication is likely to occur in all places where dust mixes with condensed water.

Although chemical treatment of raw materials and feeds can significantly reduce *Salmonella* contamination this may not be the complete solution. The concept of producing bio-secure feed in specialised and well-controlled feed mills is also an important part of the battle against pathogenic micro-organisms. Bio-secure feed mills will operate to very high standards of cleanliness and hygiene. They will have a dirty side and a clean side with good separation between the two sides. In this system manufactured feed can be produced free from contamination by micro-organisms and kept free from subsequent contamination.

Salmonella control in vehicles and silos

Use of good quality raw materials and good feed manufacturing practices are extremely important to avoid contamination of feeds by *Salmonella*. However this vigilance must be maintained by ensuring that transport vehicles and feed silos on farm are also *Salmonella*-free. Routine cleaning and disinfection of transport vehicles and silos is best done using dry or powder *Salmonella* inhibitors. Use of dry rather than water-based *Salmonella* control agents has the advantage of not wetting surfaces which is always a risk for contamination. Application of dry products is technically easy to do using powder dusters that can create a dry fog inside a truck or a silo. As the fogging is done with *Salmonella* inhibitors that are safe and edible the product used can remain in the vehicle or silo and offer continuous protection.

By judicious use of both liquid and dry *Salmonella* inhibitors it is possible to ensure that raw materials, feed manufacturing facilities, poultry feeds, transport vehicles and feed silos are kept free from *Salmonella*. These programmes clearly have to be instituted within a framework of good management and attention to other environmental factors which could introduce or disseminate *Salmonella*. Effective rodent and wild bird control is always important. Avoidance of dust in facilities is important as *Salmonella* may survive and be disseminated by dust particles.

Future directions

Improved methods for monitoring moulds, mycotoxins and pathogenic bacteria in feeds are required. There has been considerable process already in that many ELISA (enzyme linked immunosorbent assays) test kits are now available for detection of mycotoxins and of *Salmonella* antibodies in both feeds and animal tissues.

There will probably be a greater emphasis upon water activity (a_w) as an important parameter in conservation of feeds. This may allow good shelf-life of feeds to be obtained with minimum use of organic acid inhibitor products.

Better methods for removing or neutralising mycotoxins in feed raw materials are urgently needed. It will not always be possible to simply destroy contaminated feed raw materials. However given the diversity of molecular structures in the mycotoxins and their presence in small quantities, usually <10 g/tonne, this will be very difficult to achieve by chemical treatments.

Another possibility is to develop biological decontamination procedures. A variety of bacteria, yeasts, and moulds are able to degrade aflatoxins. Biological decontamination programmes are still at the research stage and their practical utility remains to be demonstrated.

The development of feeding strategies that will encourage a microflora in the gastrointestinal tract able to degrade mycotoxins could be very valuable. This probably occurs naturally in ruminants but needs to be extended to monogastric species. This will require a much greater understanding of the gastrointestinal microflora and its relationship to various feed formulations.

Control of *Campylobacter* and *Salmonella* in meat

There is considerable concern about the incidence of both *Campylobacter* and *Salmonella* in fresh meat and further research will be needed here. *Campylobacter* is a common food pathogen in poultry (Saleha, *et al.*, 1998; Atanossova and Ring, 1999) and *Salmonella* is of concern both in poultry and pig meats. However *Campylobacter* is not a pathogen of poultry and most *Salmonella* serotypes are not pathogens for pigs. Good feed hygiene will probably help control of *Salmonella* contamination but the control of *Campylobacter* is much less obvious as it seems not to be a feed-borne organism.

In Denmark a national *Salmonella* surveillance programme was implemented in 1995 (Mousing *et al.*, 1997). Large scale monitoring of pig herds for infection is carried out using a serological test. The test detects antibodies that have developed in the blood or meat juice of slaughtered pigs after infection with *S. typhimurium*. These type of programmes will be extended to other countries.

These two food pathogens are nevertheless contaminants of the carcasses of poultry or pigs. Therefore it seems likely that feed or water-based treatments will need to be developed to control these organisms. This may require several approaches using various antibacterial substances in feeds that are of natural origin such as various herbs and spices. Treatment of drinking water for broilers with acetic, formic or lactic acids may be one possibility to reduce carcass contamination by *Campylobacter* and *Salmonella* (Byrd *et al.*, 2001).

An increasing amount of animal feeds will have to be produced in bio-secure units. At the present feed from bio-secure units is usually destined for poultry breeding flocks as here the risk of contamination of birds by *Salmonella* has very serious economic consequences. Nevertheless bio-secure feed could be produced for all animals albeit at a higher production cost.

References

Adams, C. A. (1999). *Nutricines. Food Components in Health and Nutrition*. Nottingham University Press, UK. pp. 13-14.

Atanassova, V. and Ring, C. (1999). Prevalence of *Campylobacter* spp. in poultry and poultry meat in Germany. *International Journal of Food Microbiology*, **51**, 187-190.

Bailey, R. H., Kubena, L. F., Harvey, R. B., Buckley, S. A. and Rottinghaus, G. E. (1998). Efficacy of various inorganic sorbents to reduce the toxicity of aflatoxin and T-2 toxin in broiler chickens. *Poultry Science*, **77**, 1623-1630.

Borsting, C. F., Engberg, R. M., Jakobsen, K., Jensen, S. K. and Andersen, J. O. (1994). Inclusion of oxidised fish oil in mink diets. 1. The influence on nutrient digestibility and fatty-acid accumulation in tissues. *Journal of Animal Physiology and Animal Nutrition*, **72**, 132-145.

Byrd, J. A., Hargis, B. M., Caldwell, D. J., Bailey, R. H., Herron, K. L., McReynolds, J. L., Brewer, R. L., Anderson, R. C., Bischoff, K. M., Callaway, T. R. and Kubena, L. F. (2001). Effect of lactic acid administration in the drinking water during preslaughter feed withdrawal on Salmonella and Campylobacter contamination of broilers. *Poultry Science*, **80**, 278-283.

Cotton, R. T. and Wilbur, D. A. (1982). In: *Storage of Cereal Grains and Their Products*. Ed: C. M. Christensen American Association of Cereal Chemists pp. 281-318.

Davies, R. H. and Wray, C. (1996). Persistence of *Salmonella enteriditis* in poultry units and poultry food. *British Poultry Science*, **37**, 598-596.

Davies, R. H. and Wray, C. (1997). Distribution of salmonella in ten animal feedmills. *Veterinary Microbiology*, **51**, 159-169.

Duthie, G. G. (1993). Lipid peroxidation. *European Journal of Clinical Nutrition*, **47**, 759-764.

Egon Joseffson, B. G. and Möller, T. E. (1980). Heat stability of Ochratoxin A in pig products. *Journal of the Science of Food and Agriculture*, **31**, 1313-1315.

El-Banna, A. A. and Scott, P. M. (1984). Fate of mycotoxins during processing of foodstuffs. 111. Ochratoxin A during cooking of Faba bean (*Vicia faba*) and polished wheat. *Journal of Food Protection*, **47**, 189-192.

Elissade, M.H., Ziprin, R. L., Huff, W. E., Kubena, L. F. and Harvey R. B. (1994) Effect of ochratoxin A on *Salmonella*-challenged broiler chicks. *Poultry Science*, **73**, 1241-1248.

Engberg, R. M. and Borsting, C. F. (1994). Inclusion of oxidised fish oil in mink diets. 2. The influence on performance and health considering histopathological, clinical-chemical, and haematological studies. *Journal of Animal Physiology and Animal Nutrition*, **72**, 146-157.

Engberg, R. M., Lauridsen, C., Jensen, S. K. and Jakobsen, K. (1996). Inclusion of oxidized vegetable oil in broiler diets. Its influence on nutrient balance and on the antioxidative status of broilers. *Poultry Science*, **75**, 1003-1011.

Enigl, D. C., and Sorrells, K. M. (1997). Water activity and self-preserving formulas. In: *Preservative-free and self-preserving cosmetics and drugs*. Eds. J. J. Kabara and D. S. Orth. Marcel Dekker Inc., New York, USA. pp. 45-73.

Fleurat-Lessard, F. (1989). In: *Contamination par les moississures des aliments pour animaux*. Association Francaise des Techniciens de l'Alimentation Animale pp. 124-143.

Holmberg, T., Kaspersson, A., Larsson, K., and Pettersson, H. (1989). Aflatoxin production in moist barley treated with suboptimal doses of formic and propionic acids. *Acta Agricultura Scandinavica*, **39**, 457-464.

Kirby, R. M. and Davies, R. (1990). Survival of dehydrated cells of *Salmonella typhimurium* LT2 at high temperatures. *Journal of Applied Bacteriology*, **68**, 241-246.

Kubena, L. F., Harvey, W. E., Buckley, S. A., Edrington, T. S. and Rottinghaus, G. E. (1997). Individual and combined effects of moniliformin in *Fusarium fujikuori* culture material and aflatoxin in broiler chicks. *Poultry Science*, **76**, 265-270.

Lamghari, R., Villaume, C., Pelletier, X., Bau, H. M., Schwertz, A., Nicolas, J-P. and Mejean, L. (1997). Effect of rancidity of *Lupus albus* protein-concentrate-based diets on food intake and growth of wistar rats. *Journal of the Science of Food and Agriculture*, **75**, 80-86.

Li, Y. C., Ledoux, R., Bermudez, A. J., Fritsche, K. L. and Rottinghaus, G. E. (2000). The individual and combined effects of fumonisin B1 and moniliformin on performance and selected immune parameters in turkey poults. *Poultry Science*, **79**, 871-878.

Mabee, M. S. and Chipley, J. R. (1973). Tissue distribution and metabolism of aflatoxin B_1-C^{14} in broiler chickens. *Applied Microbiology*, **25**, 763-769.

Matlho, G., Sakchai, H., Riemann, H. and Kass, P. (1997). Destruction of Salmonella enteritidis in poultry feed by combination of heat and propionic acid. *Avian Diseases*, **41**, 58-61.

Mousing, J. Thode Jensen, P., Halgaard, C., Bager, F., Feld, N., Nielsen, J. P. and Bech-Nielsen, S. (1997). Nation-wide *Salmonella enterica* surveillance and control in Danish slaughter swine herds. *Preventive Veterinary Medicine*, **29**, 247-261.

Nwanguma, B. C., Achebe, A. C., Ezeanyika, L. U. S. and Eze, L. C. (1999). Toxicity of oxidised fats II Tissue levels of lipid peroxides in rats fed a thermally oxidised corn oil diet. *Food and Chemical Toxicology*, **37**, 413-416.

Ramos, A. J. and Hernandéz, E. (1997). Prevention of aflatoxicosis in farm animals by means of hydrated sodium calcium aluminosilicate addition to feedstuffs: a review. *Animal Feed Science and Technology*, **65**, 197-206.

Saleha, A. A., Mead, G. C. and Ibrahim, A. L. (1998). *Campylobacter jejeuni* in poultry production and processing in relation to public health. *World's Poultry Science Journal*, **54**, 49-58.

Schell, T. C., Lindemann, M. D., Kornegay, E. T., Blodgett, D. J. and Doerr, J. A. (1993). Effectiveness of different types of clay for reducing the detrimental effects of aflatoxin-contaminated diets on performance and serum profiles of weanling pigs. *Journal of Animal Science*, **71**, 1226-1231.

Scudamore, K. A., Nawaz, S. and Hetmanski, M.T.(1998). Mycotoxins in ingredients of animal feeding stuffs: II. determination of mycotoxins in maize and maize products. *Food Additives and Contaminants*, **15**, 30-55.

Valdivia, A. G., Martinez, A., Damián, F. J., Quezada, T., Ortiz, R., Martinez, C., Llamas, J., Rodriguez, M. L., Yamamoto, L., Jaramillo, F., Loarca-Pina, M. G. and Reyes, J. L. (2001). Efficacy of N-acetylcysteine to reduce the effects of aflatoxin B_1 intoxication in broiler chickens. *Poultry Science*, **80**, 727-743.

Van Kempen, T., Park, B., Hannon, M. and Matzat, P. (2001). Precision nutrition: weighing feed ingredients correctly. *Journal of the Science of Food and Agriculture*, **81**, 726-730.

Weidong, W., Cook, M. E. and Smalley, E. B. (1991). Decreased immune response and increased incidence of tibial dyschondroplasia caused by Fusaria grown on sterile corn. *Poultry Science*, **70**, 293-301.

3   Eating to Live: Voluntary Feed Intake

Humans are sometimes classified into two groups depending upon their cultural origins: those who eat to live and those who live to eat. As far as animal production is concerned, animals only eat to live, but it is important that they eat sufficient quantities of feed to support both life and productivity. Feed intake is one of the most important aspects of an animal's life and the amount of feed that an animal can or will consume is a factor of major economic importance and consequently is an important aspect of Total Nutrition. The largest cost in most intensive animal production systems is the feed, often amounting to around 70% of total production costs and profit from the animal production enterprise is frequently only 12-15% of total feed costs (Chadwick, 1998).

Feed intake is a vital determinant of animal growth since in general the more food that an animal consumes each day the greater will be the increase in its productivity. Increased animal growth or productivity rates obtained by increased feed intakes is usually associated with an increase in overall efficiency of animal performance. Maintenance costs are decreased proportionately as productivity rises (Table 1).

Feed intake and energy requirements by an animal are proportional to metabolic weight ($W^{0.75}$) rather than directly to liveweight. When the data are scaled to metabolic bodyweight ($W^{0.75}$) there is a clear and consistent reduction in FCR as feed intake rises indicating increased efficiency of animal performance (Lawrence and Fowler, 1997).

Table 1
Effect of growth from increasing feed intake on feed conversion ratio of pigs from 20-90 kg

Feed intake (g/kg $W^{0.75}$)	Liveweight gain (g/kg $W^{0.75}$)	FCR
73	23.8	3.07
88	33.4	2.63
102	42.9	2.38
117	52.4	2.23
132	62.0	2.12
147	71.5	2.05

Consequently feed intake is a key issue in animal husbandry and has a major influence upon health, welfare, environment and productivity of animal production (van der Heide et al., 1999).

There however certain exceptions to this generalisation. For example pregnant sows are usually restricted in feed intake in order to maintain optimal body condition and productivity. Unrestricted feeding would allow the sows to gain too much weight. Some breeds of bacon pigs would produce too much fat in the carcass if feed intake was excessive. Poultry breeding hens must be kept on a restricted feed intake regime to reach an appropriate body weight at a suitable age otherwise the quality of the hatching eggs will be poor.

Voluntary feed intake by animals is a complex issue which involves several different actions such as the search for feed, recognition of feed, the quality and palatability of the feed, the intake capacity of the animal, and the environment. The concentration of protein, the amino acid balance and the deficiency or excess of various minerals or vitamins will all influence feed intake. In addition the concentration of available energy in a diet has a major impact upon the amount of feed consumed.

Both pigs and poultry can adjust their feed intakes to compensate for changes in the concentration of available energy whether this is digestible energy (DE) for pigs or metabolisable energy (ME) for poultry. The compensation however is not exact and lower energy diets may be consumed to excess which could result in undesirable fatness or poor economic performance of the animals. Very low energy diets may not be consumed in adequate quantities and result in poor growth rates.

Animals in general seem not to vary feed intake greatly in response to variations in protein content. Very low protein contents frequently depress voluntary feed intake. For example voluntary feed intake by young pigs was depressed when the protein content of the feed was below 6% and growth was depressed by protein levels below 9% (Robinson, *et al.*, 1974).

Most animals including pigs, poultry and ruminants respond to environmental temperatures in a similar way in that high temperatures reduce feed intake and low temperatures increase feed intake. This is particularly important in the poultry industry in hot climates where maintenance of adequate feed intake is a perennial problem.

There are also important physiological differences between ruminants and monogastric animals. For example the amount of glucose absorbed from the gastrointestinal tract in ruminants is relatively small and blood glucose levels show little relationship to feed intake. The major source of energy absorbed from the gastrointestinal tract in ruminants are the volatile fatty acids, acetic, propionic and butyric and these compounds may influence feed intake in ruminants.

Different species of animals are also raised for different purposes. This can be lean tissue growth in the case of broilers and pigs, egg production in layers and milk production in dairy cows.

Physiological status and feed intake

Feed intake will also vary with physiological state of the animal. Early weaned piglets often do not eat sufficient quantities of feed for growth requirements and consequently often suffer from a post-weaning stress. Lactating animals on the other hand will have a very high feed intake compared to the non-lactating animal. Nevertheless many dairy cows and sows undergo a weight loss during early lactation as they cannot eat sufficient feed to support the quantity of milk produced. Animals suffering diseases reduce feed intake.

Reduced feed intake in response to a disease syndrome is also possibly a survival strategy evolved by the animal to facilitate recovery. Mice infected with *Listeria* and force-fed to the same energy intake as in uninfected control mice suffered much greater effects from the *Listeria* than mice which were allowed to reduce feed intake (Murray and Murray, 1979). Infected mice allowed to feed *ad libitum* ate 58% as much feed as uninfected control mice and suffered 43% mortality. Force-feeding of infected mice increased mortality from 43% to 93% and shortened the survival time of the infected mice. This strongly suggests that reduced feed intake plays an important role in defence and survival of animals when confronted with infectious diseases. It has long been a common practice in pig production to withhold feed from piglets suffering scouring problems.

Feeding in mammals and birds is also controlled by centres in the hypothalamus situated beneath the cerebrum in the brain but other areas in the central nervous system are also involved in regulating feed intake.

In poultry the capacity of the crop influences feed intake and in mammals there are receptors for satiety in the oesophagus, duodenum and small intestine. Distension due to feed consumption of these areas increases the activity of the vagus nerve and activates the satiety centre of the hypothalamus in the brain.

Hormones play an important role in controlling both feed intake or appetite and satiety where the animal is discouraged from eating. The peptide hormones, cholecystokinin and glucagon are probably the two most important hormones affecting satiety in animals. Cholecystokinin is present in the brain and is also released into the gastrointestinal tract when digestive products such as amino acids and fatty acids reach the duodenum. Systemic administration of cholecystokinin decreases feed intake in a number of mammals and it is also involved in regulating energy balance. Glucagon is a pancreatic hormone and probably also acts as a satiety factor.

The receptors involved in the satiety response to cholecystokinin are located in the stomach wall. The direct control of feeding behaviour by cholecystokinin derived from the brain is well established in sheep and pigs but not in other species (Lawrence and Fowler, 1997). It is an extremely complicated control system however as there are at least five forms of cholecystokinin present in the brain. Furthermore these variants probably have their effect mediated through the release of other brain hormones such as calcitonin and noradrenaline.

The hormone somatostatin may also play a role in this complex phenomena of feed intake and satiety. This hormone is found in both the brain and in the gastrointestinal tract. It inhibits many other peptide hormones. Other brain peptide hormones, the orexins, are also thought to play a role in stimulating feed intake but their full role is not yet established (Arch, 2000).

The presence of a number of nutrients in the blood such as glucose, fatty acids, peptides, amino acids, vitamins and minerals may cause satiety in an animal and discourage it from further consumption of feed. It has long been known that a small dose of insulin, which lowers the concentration of blood glucose, also causes an animal to feel hungry. Poultry do not seem to respond to blood glucose or other nutrient levels to the same extent as mammals.

Activation of the immune system and the subsequent production of cytokines leads to reduced feed consumption (Johnson, 1997). This is associated with increased body temperature and reduced growth. The cytokine signal from the immune system allows the body to re-direct its metabolic activities and diverts nutrients away from growth towards fighting the disease challenge. This reduction in growth due to activation of the immune system is now seen as a major obstacle to growth efficiency and is important in the raising of animals for food where maximum growth is required. Diets which do not activate the immune system and reduce feed intake will be more advantageous in terms of the efficiency of animal production (Chapter 6).

Pigs

A major source of economic loss for pig producers is the difficulty of maintaining good growth of piglets immediately after weaning. Piglets suffer a post-weaning stress due to all the changes of environment, absence of the sow, introduction to new feed based on plant carbohydrates instead of lactose in sow milk. The major driver for post-weaning growth is feed intake in piglets but this would be further reduced by intake capacity if low energy diets were fed. However in piglets it seems that voluntary feed intake is less than it could be for up to three weeks post-weaning and this has a subsequent effect upon performance in the fattening stage. In practice many weaned pigs may eat less than their maintenance requirement for some three days or more post-weaning. This reduced feed intake is also associated with atrophy of the villi in the gastrointestinal tract and this in turn leads to various gastrointestinal disorders (Pluske *et al.*, 1996). A continuous supply of nutrients is crucial for the maintenance of the integrity of the gastrointestinal tract and of the digestive capacity.

In young weaned pigs feed intake is influenced by health, space, environment, gut capacity, feed type, age and weight at weaning. The young pig has a high potential for growth but this is frequently unrealized due to insufficient energy intake. Whenever the weaning age is less than 32 days there is a serious need to consider feeding and health. In physiological terms probably weaning at 30-32 days is most effective.

The growth performance of the young pig from 5-25 kg liveweight is limited also by the physical ability to take in nutrients. Consequently processes such as concentration of the diet which can increase nutrient intake may have a beneficial effect upon overall early growth performance.

Feed intake is inevitably curtailed by limitations of gut capacity on diets of less than 15 MJ DE/kg (Whittemore and Green, 2001). To maintain a pre-weaning growth rate of some 280 g/day a 6 kg piglet would need to eat 475g of a nutrient dense (16.5 MJ DE/kg) feed (Fowler and Gill, 1989). In actual reality weaned piglets may eat less than their maintenance requirement for some three or more days post-weaning. Maintenance of uninterrupted growth patterns after weaning can only be achieved under very high standards of husbandry and nutrition.

Lactating sows present a particular problem of feed intake as their nutritional requirements are not usually satisfied from dietary sources as their feed intake is not sufficient. Consequently sows mobilize their body reserves and this may result in impairment of their subsequent reproductive performance. This is a major challenge in Total Nutrition to ensure that sows are adequately fed and new diets will need to be formulated with this objective in mind. The production of higher density sow diets by adding extra fat is not a solution under temperate climatic conditions (Christont *et al.*, 1999). Sows fed increased levels of dietary fat decreased feed intake so that metabolizable energy intake remained constant.

Poultry

Poultry seem to have a good ability to modulate their feed intake to satisfy the requirements for energy, calcium, phosphorus, zinc, thiamin and various amino acids. This is utlized in practice by feeding whole cereal grains such as whole wheat together with a manufactured broiler feed. In this system, increasing amounts of whole wheat are added into a broiler feed, and the birds are able to select appropriate quantities of the whole wheat grains and of the manufactured feed to satisfy their nutritional requirements. This system is usually cheaper than feeding only a manufactured feed because milling, mixing and pelleting costs are avoided for the quantity of whole cereal incorporated in the feed.

Ruminants

Feed intake in ruminants, is extremely important in maintaining good production levels and feed intake is a primary constraint on milk yield for high producing dairy cows. Feed intake is complex and has been the focus of much attention over many years. It is influenced by numerous factors relating to the diet and the environment. A substantial reduction in feed intake is frequently observed in late pregnancy and this may continue into early lactation. This feed intake reduction has traditionally been ascribed to physical constraints on the volume of feed that could be consumed. However this feed intake reduction also coincides with changes in reproductive status, fat mass, and metabolic changes in support of lactation.

Ruminants differ from monogastric animals in the way they digest feed. Rumen micro-organisms ferment feed to produce short chain volatile fatty acids that are the main energy source, generally accounting for 50-75% of energy digested. Their production through fermentation increases very rapidly when feed arrives in the rumen and it is not surprising that they should play an important role in controlling feed intake.

It remains an attractive but elusive goal to be able to regulate the feed intake in dairy cows. Increased feed intake will improve production efficiency and animal health. This is of particular interest in the concept of Total Nutrition where nutrition has to be directed towards improving health as well as economic productivity. Many of the health problems in the high performing dairy cow, both metabolic or non-infectious and infectious diseases occur in early lactation and are probably related to reduced feed intakes just prior to parturition. Infectious diseases which lead to activation of the immune system and the production of cytokines also generally reduce feed intake (Ingvartsen and Andersen, 2000).

In ruminants feed intake is largely controlled by signals generated from the presence of feed in the rumen. The distension of the rumen and the various chemical or biochemical signals triggered by rumen digestion enable ruminants to control their short term feed intake. Over the longer term, ruminants seem able to select feed both to optimise the functioning of the rumen and to satisfy the nutritional equilibrium the animal requires.

Infusion of a mixture of volatile fatty acids into the rumen causes a decrease in feed intake during the course of a meal (Faverdin, 1999). Feed intake is inversely proportional to the amount of volatile fatty acids infused into the rumen. This satiation response triggered by volatile fatty acids makes it possible for the animal to evaluate the size of a meal and to prevent excess consumption that could be detrimental to the proper functioning of the rumen or even to the health of the animal.

Ruminants seem to detect and develop a preference for sweetened diets. For example cows showed a preference for feeds which contained sucrose (Nombekela et al., 1994). Intestinal sugars actually represent a relatively minor energy source in ruminant nutrition so this is somewhat surprising. However this preference for sweetness is not of great practical significance as it does not enable feed intake to be increased over the long term (Nombekela and Murphy, 1995).

Many shrub plants used for ruminants have high levels of tannins and these suppress feed intake. Tannins are able to bind to proteins, cell walls and inhibit rumen microbial and digestive enzyme activities. Polyethylene glycol (PEG) complexes with tannins and addition to lamb diets encouraged consumption of feeds with high tannin levels (Titus, 2000). Polyethylene glycol is a non-toxic compound and so can safely be fed to animals in cases where high tannin feeds must be used.

Water intake

Access to high quality safe drinking water is clearly as essential for health and growth of animals as it is for humans. Water is actually the single most important nutrient, as a lack of water will cause illness and death more rapidly than the lack of any other nutrient. Unfortunately water often is the "forgotten nutrient" even though for most animals the water intake is 1.5-2.0 times the weight of feed consumed (Table 2) (Cobb, 2000). Indeed in many operations a measure of the water:feed intake ratio is a useful management tool to indicate health and performance status of the animals. Reduced water supply inevitably leads to a reduction in feed consumption because many of the functions of water are involved in the digestion and metabolism of feed. Monogastric animals usually drink before, during, or after each meal, whereas ruminants with their great capacity for storing digesta with a high water content often eat feed without drinking, but then when they do drink they take in a very large volume of water.

The whole process of feed digestion and absorption takes place essentially in a high-moisture environment in the gastrointestinal tract.

Age	Consumption by 1000 broilers/day	
(days)	Feed (kg)	Water (litres)
7	27	58-65
14	59	102-115
21	99	149-167
28	135	192-216
35	165	232-261
42	181	274-308

Table 2
Weight of water
and feed
consumed by
broilers

Therefore even though feed may be consumed with a moisture content of 11-14% it must be rapidly hydrated in the gastrointestinal tract for efficient utilization and clearly adequate supplies of drinking water will be essential to maintain efficient feed utilization. Water is also a major constituent of any living organism and is an important medium in animals for excretion of soluble waste products in the urine.

Water is also a medium for the growth and dissemination of pathogenic micro-organisms and so the quality of drinking water is extremely important. This becomes even more critical when there is less recourse to various antibiotics and other drugs. The drinking water supply is already routinely used to deliver vaccines to animals, particularly broilers. It may be possible to develop programmes with various organic acid mixtures in the drinking water to combat food pathogens such as *Campylobacter* in broilers (Byrd *et al.*, 2001).

Environment and feed intake

The most important environmental factors that affect feed intake are temperature, housing and social conditions and the incidence of diseases.

Environmental temperature markedly influences voluntary feed intake. Body heat is produced in an animal during the digestion and metabolism of feed and this heat production could also be a signal used to regulate feed intake. At high temperatures body temperature rises and feed intake declines in an attempt to reduce the heat production associated with feeding, digestion, absorption and metabolism and to moderate any further increase in body temperature. It is well known that in hot climates poultry in particular reduce feed consumption, sometimes to undesirably low levels. By contrast in many species feed intake increases in cold environments. Cold temperatures require the animal to increase its rate of heat production in order to maintain the appropriate body temperature. This is an energy demanding process and so usually feed intake will increase.

Housing and social conditions, particularly for weaned piglets play a very important role in affecting feed intake. Good housing protects animals from extremes of temperature, rain and wind all of which can affect feed intake. Social conditions such as stocking density and availability of feeding and drinking places are obviously also important management factors in affecting voluntary feed intake.

In all animals a reduction in feed intake is one of the first signs of disease. This is related to the activity of the immune system when proteins known as cytokines are produced that increase body temperature and reduce appetite. This is further discussed in Chapter 6.

Prediction of voluntary feed intake

The prediction of voluntary feed intake by animals based on some parameter of the feed would be a very useful tool to optimise the economic performance of animal production. Such a tool would allow the production of feed formulations to ensure optimal nutrient intake. It could also be used to produce formulations to reduce feed intake when necessary such as for gestating sows or breeding hens.

Intensive animal production systems are able to control closely the environment of the animals. The supply of feed can be regulated and assured, diet formulations can be controlled, external risks such as predation or diseases can be avoided, movement can be constrained, space and social conditions can be manipulated, environmental temperatures can be controlled. Nevertheless the problem of feed intake prediction is still extremely complicated because it must encompass different species of animal, with varying feed composition and varying environments. Feed composition will frequently vary in contents of protein, fat and carbohydrates, digestibility of proteins and fibres and the moisture content. The climate will also seriously influence feed intake.

Many animals raised in controlled environments without limitations on feed supply seem to be able to adjust their feed intake to the energy value of the feed so that they consume a constant amount of calories. This is sometimes described as "eating to requirements" and the total voluntary feed intake of an animal is a combination of two factors; maintenance requirement and growth requirement. In intensive production where feed is unlimited and environment is adequately controlled this model of feed intake has been remarkably successful in predicting feed intake of growing and reproducing animals (Yearsley *et al.*, 2001). The theory of this model assumes that an animal eats sufficient feed in order to support its genetic growth rate or productivity. This further assumes that constraints such as gut volume and environment are not limiting factors. This has been frequently demonstrated in poultry where offering a low energy diet will increase the weight of feed consumed. If a diet is too low in energy content however animals may not be able to eat sufficient to obtain enough energy because other factors such as gastrointestinal capacity become limiting and predictive models will not accurately reflect the real situation.

Predicting feed intake in pigs is difficult because it is influenced by DE content, amount of indigestible matter and feed bulk. An interesting concept, the water holding capacity (WHC) of feeds may be a useful parameter to predict the intake of fibrous feeds in pigs (Tsaras, *et al.*, 1998). The WHC of a feed is a measure if its ability to immobilize water within the feed matrix. The WHC depends upon the presence of various polysaccharides which trap water, swell and form bulky gels of high water content. The WHC concept has the advantage that it can be rapidly, cheaply and accurately measured on various feed raw materials and feeds.

Most animals will increase feed intake when a high quality feed is progressively diluted with poorly digestible materials which in practice would usually be high in dietary fibre. There will be a point however when the animal will fail to increase intake to compensate fully for an excessively high fibre content. This might be due to the physical bulk of the feed exceeding the capacity of the gastrointestinal tract. This will also vary due to the age of the animal. Mature animals, particularly pigs or ruminants could more readily increase feed intake than could young piglets or calves or fast growing broilers.

Feed palatability and sensory appraisal

In humans the senses of sight, smell, touch and taste play an important role in stimulating appetite and influence the quantity of food consumed. However it cannot be assumed that animals utilise these same sensory characteristics towards feed. A good example is colour which is extremely important in humans in indicating the acceptance of food whereas in animals feed colour has little impact. Many animal species have poor colour vision and so feed colour is unlikely to be as important as it is for humans. Indeed ruminants will graze in poor light conditions and even eat in darkness where colours could not be detected. It seems that in general animals are less influenced by colour and taste of feed than are humans. In some mammals however, the sense of smell is very well developed and much more sensitive than in humans. Dogs for example have a much more powerful sense of smell than humans yet are far less discriminating in their feeding habits. There is nevertheless a great tendency to assume that animals have the same attitudes to feed as humans have to food.

It is difficult to establish true feed palatability characteristics for animals. In many cases the taste preferences of humans are used as the initial position but this is not necessarily reflected in animal responses. Various "preference tests" are widely used in animal nutrition as experimental procedures and give useful results. They can however always be criticised on the basis that farm animals are not usually given a choice of feeds but only one single feed that they either eat or do not eat.

Most animals in fact will show preference for certain feeds when given a choice. Piglets and calves seem to prefer sweetened feeds to unsweetened feeds. Poultry are indifferent to the common sugars but find xylose objectionable. Feeding trials based on choice of two feeds is also difficult to interpret because inevitably in any group of animals there is a degree of variability. In choice feeding studies it is very rare to have a result of 100% to 0% with two feeds. It is much more common that a small percentage of the animals prefer one feed and a larger percentage prefer the other feed. It is far from clear why even a small percentage of test animals should prefer a different feed from its companions in a group. It is difficult to translate such observations into practical situations as animals do not normally receive a choice of feeds but are usually confronted with only one feed.

Nevertheless preference testing has been widely used and gives some useful indications of the response of animals to various flavours. As a result of these tests it has been established that pigs exhibit a strong preference for aqueous solutions of sucrose, with some preference for lactose and sodium saccharin compared to water alone. Curiously enough sodium cyclamate was not preferred (Glaser *et al.*, 2000).

Furthermore it has also been demonstrated by electophysiological measurements that several other compounds tasting sweet to humans such as monellin, thaumatin, or aspartame do not elicit any significant response in pigs. It was therefore concluded that these compounds do not taste sweet to pigs.

It appears that sucrose is the most preferred sweetener in pigs. Sucrose and fructose gave similar responses in pigs and humans but lactose, maltose, glucose, and galactose are two times less efficient in pigs. Out of a range of intense sweeteners tested, several were ineffective (aspartame, cyclamate, monellin, neo-hesperidine dihydrochalcone, and thaumatin). However acesulfame-K, saccharin, dulcin and sucralose were quite well received by pigs as sweeteners.

Grazing ruminants seem to prefer young green forage to dry or old material and they prefer leaf to stem. Colour and appearance of feed seem unimportant in ruminants as they will eat in complete darkness. There is concern about the acceptability of mixed feed for high–producing dairy cows where large quantities of feed need to be consumed to support high levels of lactation. Frequently various flavours are incorporated into mixed dairy feeds or molasses may be used as both an energy source and a palatability enhancer.

In sheep both glucose and orange elicited positive flavour responses (Ralphs *et al.*, 1995). However the glucose response may be more complicated than simply a flavour preference since the high glucose feeds increased the rumen microbial mass. This in turn may have generated nutrients for further absorption by the animal in the lower part of the gastrointestinal tract and could act as a positive feedback system to improve feed intake rather than being a direct response to flavour.

Wheat straw flavoured with onion or oregano together with sodium propionate was also well accepted by sheep in preference trials (Villalba and Provenza, 1996). However flavoured feeds with sodium chloride instead of sodium propionate were strongly avoided by the sheep.

A comparison of feeds for lactating cows flavoured with the four basic flavours sweet (sucrose), sour (HCl), bitter (urea) and salty (NaCl) showed that sucrose was the preferred flavour (Nombekela *et al.*, 1994). However the unflavoured control was the second preference and so clearly the cows could detect and discriminate between various flavours in feeds. In another experiment where anise, monosodium glutamate, dehydrated alfalfa meal flavour and molasses flavour were tested. The control feed and the monosodium glutamate ranked equally first. Of all the flavours tested only sucrose seemed to be able to increase feed intake.

However in subsequent work the feed consumption of dairy cows was not enhanced by sucrose supplementation. Results suggested that sucrose might have transiently increased feed intake in the first two weeks postpartum (Nombekela and Murphy, 1995). It is frequently observed that feed intakes of test animals return to normal amounts soon after initial increases when given flavoured feeds. This makes it extremely difficult to ascertain that a feed flavour will permanently induce an increase in feed intake.

It is clearly difficult in practice to improve voluntary feed intake feed permanently by utilization of various flavours and sweeteners that appeal to humans. Nevertheless even though it may be difficult to improve palatability of feeds it is relatively simple to reduce palatability. The problems of the deterioration feed quality due to moulds and oxidation discussed in Chapter 2 is extremely relevant in terms of palatability.

Oxidised or rancid feeds are likely to be consumed by animals at a reduced rate which affects growth and performance, and this has been demonstrated in rats (Lamghari *et al.*, 1997). A lupin-based feed was stored for 15 days and then heated which oxidised the unsaturated fatty acids in the feed. The rats showed significantly reduced feed intake and body weight gain on the oxidised diet compared to a diet which was not oxidised.

Similarly with microbial spoilage it is widely recognised that mouldy feeds have reduced palatability. In pigs this can be traced to a particular mycotoxin, deoxynivalenol (DON) also known as "vomitoxin" which causes serious feed refusal and reduced intake (Table 3). It was calculated that feed intake is depressed by 6% for every ppm of DON in the feed (Blaney and Williams, 1991).

Table 3
Effect of deoxynivalenol (DON) on pig performance from 20-50 kg

Performance characteristic	Deoxynivalenol levels in feed (ppm)			
	0	*4*	*8*	*11*
Average daily feed intake (kg)	2.05	1.76	1.47	1.28
Average daily gain (kg)	0.89	0.80	0.58	0.45
Feed/gain ratio	2.32	2.22	2.63	2.94

When pigs were fed a diet based on *Fusarium*-infected maize feed intake dropped from 1.38 kg/day to 1.03 kg/day over 29 days (Williams *et al.*, 1994). This drop in feed intake could not be overcome by the addition of sweeteners to the feed.

Probably the most powerful influences upon voluntary feed intake are exerted by general feed quality characteristics. The measures outlined in Chapter 2 to protect feeds against the destructive processes of autoxidation and mould growth are also the measures most likely to influence feed intake in a positive manner. Whilst antioxidants and mould inhibitors are not usually though of as palatability agents in reality they probably exert a greater influence upon feed intake than do the conventional flavours and sweeteners. Indeed if feed is rancid or mouldy addition of flavours is unlikely to be able to overcome these problems.

Appetite

Appetite is of major concern in pigs and ruminants and especially so during lactation. The demands for adequate milk production at this time require a very substantial increase in feed intake. If appetite is not sufficient to support lactation then loss of body weight will occur as the maternal body tissue is catabolized to support lactation. In pigs the required feed intake during lactation is about three times the level needed in pregnancy. It, is also possible that with the selection of growing pigs for the efficient production of lean meat, appetite has been reduced and this will be particularly significant for the lactating sow (Cole, 1990).

During lactation temperature is the most important environmental factor influencing voluntary feed intake in pigs. High temperatures due to poorly designed housing or tropical climates reduce feed intake. In general sows reduce feed intake by 100-300 g/day for every 1°C rise in temperature over the range of 16-30°C (Close and Cole, 2000).

Problems of reduced feed intake as a result of high environmental temperature can be ameliorated through appropriate feed formulation. High nutrient density diets and particularly high fat diets are useful here. Fat has a lower heat increment of feeding than has carbohydrates or protein, which allows higher energy consumption and to a large extent compensates for the reduction in voluntary feed intake at high temperatures. However when high levels of fat are used this increases the danger of oxidation and the development of rancidity in the feed which will lead to reduced palatability of the feed. Consequently good antioxidant protection of fats will be important to maintain feed palatability and support adequate feed intake (Chapter 2).

Important factors to enhance voluntary feed intake

Total Nutrition is a strategy to raise animals with minimal recourse to medication and to obtain the maximum benefit from the diet to support both growth and health. These objectives can only be achieved if the target animals consume adequate quantities of feed and therefore voluntary feed intake is a crucial element in Total Nutrition. The following practical points are important in helping to ensure an adequate voluntary feed intake by animals raised for food:

- Feed must be palatable
- Feed must be well-balanced with suitable nutrient specifications
- Ensure that feed is fresh and clean and not stale or dirty
- Protect high fat feed raw materials and feeds against oxidation
- Protect feed raw materials and feeds against moulds and mycotoxins
- Ensure that clean fresh water is freely available
- Pelleted or meal feeds should be low in dust or fines

Future directions

Further studies in feeding behaviour and feed intake will be important in the future. Understanding the effects of infections and disease on feed intake will be difficult and requires some new ideas. Feeding and issues of animal welfare will remain major concerns for consumers and legislators as well as for the animal production industry.

Restricted feeding of pregnant sows is recognised as an important factor in the development of stereotypic behaviour in sows and this is undesirable from the welfare point of view. There needs to be further research to develop appropriate feeding regimes which do not allow sows to gain excessive amounts of weight and also do not predispose them to stereotypic behaviour.

Feeding sows on a high-fibre diet containing sugar beet pulp was successful in maintaining sow condition and productivity. However the relatively large amount of feed consumed (4.1 kg/sow per day) would significantly increase the operational costs of a large sow unit (Whittaker *et al.,* 2000), so simply feeding low density feeds does not necessarily solve the problem.

Reduced feed intake brought about by the production of cytokines in response to a stimulation of the immune system may needs much further investigation. It is well established that stimulation of the immune system in pigs definitely reduces growth (Stahly, 1996). However, this may also be a protective mechanism in sick animals and may be a specific strategy to facilitate recovery and survival of the animal rather than an unfortunate response to the disease. There is clearly an immune paradox operating where reduced feed intake may help an animal recover from an infectious disease but this will reduce growth rates. The concern in Total Nutrition is to maintain high feed intakes and good animal health and welfare. This requires a better understanding of the mechanisms regulating feed intake and metabolism and their integration.

Feed flavour components that are based on various plant extracts will play an increasingly important role in Total Nutrition as they have many other biological activities in addition to a flavour characteristic. For example rosemary extracts have a powerful antioxidant activity, whereas thyme and oregano have antimicrobial activities. Many botanical extracts may well influence the immune system and function as immunomodulators. The great chemical complexity of feed flavours also necessitates relatively light regulatory controls and this will encourage the use of so-called "flavours" in more functional roles. As flavours are based on well tried and tested components largely of natural origin this development should be quite appropriate for the concept of Total Nutrition where feeds are required to maintain health as well as supply nutrients.

References

Arch, J. (2000). Orexins, feeding the big picture. *British Journal of Nutrition,* **84**, 401-403.

Blaney, B. J. and Williams, K. C. (1991). Effective use in livestock feed of mouldy and weather-damaged grain containing mycotoxins-case histories and economic assessment pertaining to pig and poultry industries of Queensland. *Australian Journal of Agricultural Research*, **42**, 993-1012.

Byrd, J. A., Hargis, B. M., Caldwell, D. J., Bailey, R. H., Herron, K. L., McReynolds, J. L., Brewer, R. L., Anderson, R. C., Bischoff, K. M., Callaway, T. R. and Kubena, L. F. (2001). Effect of lactic acid administration in the drinking water during preslaughter feed withdrawal on Salmonella and Campylobacter contamination of broilers. *Poultry Science*, **80**, 278-283.

Chadwick, L. (1998). *The Farm Management Handbook 1998/99*. Edinburgh, Scottish Agricultural College, UK.

Christont, R., Saminadin, G., Lionet, H. and Racon, B. (1999). Dietary fat and climate alter food intake, performance of lactating sows and their litters and fatty acid composition of milk. *Animal Science*, **69**, 353-365.

Close, W. H. and Cole, D. J. A. (2000). *Nutrition of Sows and Boars*. Nottngham University Press, UK, pp.181-204.

COBB, (2000). Commercial literature

Cole, D. J. A. (1990). Nutritional strategies to optimize reproduction in pigs. *Journal of Reproduction and Fertility*, Supplement **40**, 67-82.

Faverdin, P. (1999). The effect of nutrients on feed intake in ruminants. *Proceedings of the Nutrition Society*, **58**, 523-531.

Fowler, V. R. and Gill, B. P. (2000). Voluntary food intake in the young pig. In: *The Voluntary Feed Intake of Pigs, BSAS Occasional Publication No. 13*. Eds: J. M. Forbes, M. A. Varley and T. L. J. Lawrence pp. 51-60 and pp.213-215.

Glaser, D., Wanner, M., Tinti, J. M. and Nofre, C. (2000). Gustatory responses of pigs to various natural and artificial compounds known to be sweet in man. *Food Chemistry*, **68**, 375-385.

Ingvartsen, K. L. and Andersen, J. B. (2000). Integration of metabolism and intake regulation: A review focusing on periparturient animals. *Journal of Dairy Science*, **83**, 1573-1597.

Johnson, R. W. (1997). Inhibition of growth by pro-inflammatory cytokines: an integrated view. *Journal of Animal Science*, **75**, 1244-1255.

Lamghari, R., Villaume, C., Pelletier, X., Bau, H. M., Schwertz, A., Nicolas, J-P. and Mejean, L. (1997). Effect of rancidity of *Lupus albus* protein-concentrate-based diets on food intake and growth of wistar rats. *Journal of the Science of Food and Agriculture*, **75**, 80-86.

Lawrence, T. L. J. and Fowler, V. R. (1997). *Growth of Farm Animals*, CAB International, UK pp.120-121.

Murray, M. J. and Murray, A. B. (1979). Anorexia of infection as a mechanism of host-defence. *American Journal of Clinical Nutrition*, **32**, 593-596.

Nombekela, S. W., Murphy, M. R., Gonyou, H. W. and Marden, J. I. (1994). Dietary preferences in early lactation cows as affected by primary tastes and some common feed flavours. *Journal of Dairy Science*, **77**, 2393-2399.

Nombekela, S. W. and Murphy, M. R. (1995). Sucrose supplementation and feed intake of dairy cows in early lactation. *Journal of Dairy Science*, **78**, 880-885.

Pluske, J. R., Williams, I. H. and Aherne, F. X. (1996). Maintenance of villus height and crypt depth in piglets by providing continuous nutrition after weaning. *Journal of Animal Science*, **62**, 131-144.

Ralphs, M. H., Provenza, F. D., Wiedmeier, R. D. and Bunderson, F. B. (1995). Effects of energy source and food flavour on conditioned preferences in sheep. *Journal of Animal Science*, **73**, 1651-1657.

Robinson, D. W., Holmes, J. H. G. and Bayley, H. S. (1974). Food intake regulation in pigs 1. The relationship between dietary protein concentration, food intake and plasma amino acids. *British Veterinary Journal*, **130**, 361-365.

Stahly, T. (1996). Impact of immune system activation on growth and optimal dietary regimens of pigs. In: *Recent Advances in Animal Nutrition*. Eds: P. C. Garnsworthy, J. Wiseman and W. Haresign. Nottingham University Press, UK. pp. 197-206.

Tsaras, L. N., Kyriazakis, I. and Emmans, G. C. (1998). The prediction of voluntary food intake of pigs on poor quality foods. *Animal Science*, **66**, 713-723.

Titus, C. H., Provenza, F. D., Perevolotsky, A. and Silanikove, N. (2000). Preferences for foods varying in macronutrients and tannins by lambs supplemented with polyethylelene glycol. *Journal of Animal Science*, **78**, 1443-1449.

Van der Heide, D., Huisman, E. A., Kanis, E., Osse, J. W. M. and Verstegen, M. (1999). *Regulation of Feed Intake*. CABI Publishing, UK.

Villalba, J. J., and Provenza, F. D. (1996). Preference for flavored wheat straw by lambs conditioned with intraruminal administration of sodium propionate. *Journal of Animal Science*, **74**, 2362-2368.

Whittaker, X., Edwards, S. A., Spoolder, H. A. M., Corning, S. and Lawrence, A. B. (2000). The performance of group-housed sows offered a high fibre diet *ad libitum*. *Animal Science*, **70**, 85-93.

Whittemore, C. T. and Green, D. M. (2000). Growth of the weaned pig. In: *The Weaner Pig*. Eds: M. A. Varley and J. Wiseman. CABI Publishing, Wallingford, UK. pp. 1-17.

Williams, K. C., Blaney, B. J. and Peters, R. T. (1994). Pigs fed *Fusarium*-infected maize containing zearalenone and nivalenol with sweeteners and bentonite. *Livestock Production Science*, **39**, 275-281.

Yearsley, J., Tolkamp, B. J. and Ilius, A. W. (2001). Theoretical develoments in the study and prediction of food intake. *Proceedings of the Nutrition Society*, **60**, 145-156.

4   Raw Material Processing: Digestion and Absorption of Nutrients

Modern animal feeds whether for ruminants or monogastric species are largely based on raw materials derived from plants. In particular starch from cereal grains, protein and oils from oilseeds and various forages are the basis of modern diets used in animal agriculture. Relatively small quantities of raw materials of animal origin may also be used such, as meat and bone meals and fishmeals but the use of these materials is now declining in the wake of the BSE disease in cattle. Many ingredients in feed such as starch, protein or fats (triacylglycerides) are large molecules which have to be broken down or digested into simpler compounds before they can be absorbed across the mucosal membrane of the gastrointestinal tract into the bloodstream. It is only nutrients absorbed across the intestinal wall, after digestion, that are of benefit to the animal for maintenance of good health and for growth.

Digestion occurs in the gastrointestinal tract which is the largest organ in the body and it represents an interface of the body between the external environment and the interior of the body. Feed, produced external to the body, is a complex mixture of many chemical species and micro-organisms and is introduced into the gastrointestinal tract for digestion and absorption. Many different physical and chemical processes occur in the gastrointestinal tract including, comminution and mixing of ingested feed, transport of ingested materials along the gastrointestinal tract, enzymatic digestion of feed ingredients, absorption of nutrients into the body and the elimination of undigested residues.

The main site of nutrient digestion and absorption is the small intestine and when the digested feed mixture has reached the entrance to the large intestine most of the hydrolysed nutrients should have been absorbed. Indeed the capacity of the small intestine to absorb digested nutrients such as amino acids and monosaccharides may be a limiting factor for growth of the animal. This may be important for modern livestock production where high density diets are used and large quantities of nutrients need to be absorbed from the small intestine to sustain animal performance and economic competitiveness. Development of the full capacity of this nutrient delivery system of both ruminant and monogastric species is extremely important to allow full expression of genetic potential for production of the modern animal breeds.

A major challenge in future developments of animal nutrition will be to use ever more cost-effective feed ingredients and at the same time to ensure that these are efficiently digested and absorbed by the animal.

This will require careful processing of feeds and application of various nutricines to assist and enhance digestion and absorption.

Digestion

Digestion involves physical, chemical and microbial processes. The physical processes are chewing and mastication and the muscular contractions of the gastrointestinal tract which thoroughly mix and emulsify feed ingredients with various bile salts, phospholipids and enzymes secreted into the gastrointestinal tract. Muscular contractions of the circular muscle of the wall of the gastrointestinal tract or peristalsis, move the intestinal contents along the gastrointestinal tract. The chemical processes are largely hydrolytic reactions carried out by various digestive enzymes produced by the animal. Exogenous enzymes are now frequently added to animal feeds and these also play a role in digestion of feed ingredients. Microbial digestion of feed is of major importance in ruminants and is brought about by the action of the rumen microflora largely bacteria and protozoa. In monogastric animals, some microbial digestion or fermentation occurs in the large intestine and this is quite significant in pigs.

Digestion begins in the mouth and is completed when nutrients in the feed are absorbed across the villi of the small intestine or when undigested residues are eliminated from the body as faeces. In the pig, α-amylase is present in saliva and this may initiate digestion of starch immediately feed is consumed and before it enters the first part of the digestive tract, the stomach.

The stomach of monogastric mammals is a large storage organ as well as a site for digestion of feed. The stomach secretes a gastric juice which contains proenzymes, the pepsinogens, inorganic salts, mucus, and hydrochloric acid. Pepsinogens are the inactive form of pepsins which are protein-hydrolysing enzymes. The acid concentration of the gastric juice is sufficient to drop the pH of the stomach to around 2.0-3.0 and the acid activates pepsinogens converting them into pepsins. The stomach is a major site for protein digestion by the pepsin enzymes which are active at the low pH of the stomach. Pepsins preferentially attack those peptide bonds adjacent to the aromatic amino acids, phenylalanine, tryptophan and tyrosine. Pepsins also have a strong clotting action on milk. In the stomach proteins are broken down mainly into polypeptides and oligopeptides with a limited production of amino acids.

The partially digested feed or chyme leaves the stomach and it enters the small intestine which comprises the duodenum, the jejunum and the ileum. Most of the digestion takes place in the duodenum. The small intestine is also the main absorption site for digested nutrients.

The inner surface of the small intestine is lined by a series of finger-like projections, the villi, which greatly increase the surface area available for absorption of nutrients.

The feed mixture or chyme, leaving the stomach has an acid pH of 2-3 and is mixed with secretions from the duodenum, liver and pancreas. These various secretions act as a lubricant and have an alkaline pH to buffer the hydrochloric acid produced in the stomach which protects the wall of the duodenum. The pH of the chyme rises to 6-7 which is more suitable for extensive hydrolysis of the feed ingredients by the extensive range of peptidases, carbohydrases and lipases, produced by the pancreas and secreted by the cells of the villi.

The pancreas produces a very complex mixture of digestive enzymes which are secreted into the duodenum. The hormone cholecystokinin plays an important role here as it stimulates the secretion into the pancreatic juice of proteolytic proenzymes such as trypsinogen, chymotrypsinogen, procarboxypeptidases A and B, proelastase. These are all converted into active forms of trypsin, chymotrypsin, carboxyeptidase and elastase and digest dietary proteins. Trypsin and chymotrypsin hydrolyse protein molecules that were not hydrolysed by pepsin in the stomach. The pancreatic juice also contains α-amylase which digests starch, lipase which hydrolyses fats to monoglycerides, phospholipases which hydrolyse lecithins and nucleases which digest various nucleic acids. Aminopeptidases attack the peptide bond from the free amino end of peptides and dipeptidases convert dipeptides into free amino acids.

Protein digestion is of major importance in animal nutrition as animals are totally dependent upon ingested proteins for their nitrogen source. Whilst animals can utilize free amino acids such as lysine and methionine these are not found free in feed ingredients. Consequently a great diversity of proteolytic enzymes must be produced in substantial quantities. The inherent susceptibility of various feed proteins to enzymatic degradation differs and also exerts an effect on protein utilization. Highly insoluble or severely heat-denatured proteins will have reduced digestibility even though there may be an excess of proteolytic enzymes present. This limits the use of some protein sources such as feather meal for example. It is important in Total Nutrition that the inherent digestibility of sources of feed proteins is taken into account.

In cattle, pigs and poultry fed on cereals considerable quantities of highly digestible starch enter the small intestine as the major dietary energy source. α-amylase secreted in saliva and pancreatic fluid partially hydrolyses the starch granules into a range of oligosaccharides. Starch digestion is complicated by the fact that starches from different plant sources have different digestibility for animals. For example maize starch is very readily digested by poultry but less so by ruminants. Raw potato starch is poorly digestible by almost all animals.

The disaccharides sucrose and lactose may occur as feed ingredients and these, together with oligosaccharides from starch digestion are then further hydrolysed to monosaccharides by the action of a range of disaccharidases produced by the villi which line the walls of the small intestine. These enzymes include sucrase which converts sucrose into glucose and fructose, maltase which converts maltose into glucose, and lactase which converts lactose into glucose and galactose.

In modern rations for poultry and pigs fat is a major energy source and is essential to build high density diets for the modern broiler. Fat digestion is a fundamental problem because fats are water insoluble and digestion occurs in an aqueous system. An essential first step in fat digestion is to emulsify the fats and bring them into an aqueous medium with the lipases. This is brought about by bile fluids which are secreted by the liver and enter the duodenum through the bile duct. Bile is a complex fluid containing salts of the bile acids, glycocholic and taurocholic acids, bile pigments biliverdin and bilirubin, cholesterol, phospholipids and mucin. Bile fluid plays an important role in digestion and absorption of nutrients by activating pancreatic lipase, and emulsifying fats. The surfactants in the bile salts also generate microdroplets or micelles, where monoglycerides and free fatty acids aggregate together into three-dimensional structures. The micelles are transported to the wall of the gastrointestinal tract and deliver nutrients to the villi lining the wall of the gastrointestinal tract for absorption. Some of the bile acids pass through the remainder of the gastrointestinal tract and are excreted in the faeces. A portion of the bile acids is also absorbed by the ileum and returned to the liver for re-use in the digestive process.

Lysophospholipids also play a useful role in micelle formation as they are more water dispersible than other phospholipids. These may be particularly important in helping with nutrient absorption in young animals.

In all diets there is always a certain amount of feed material which is resistant to the digestive enzymes in the small intestine and a large amount of this is described as "dietary fibre".

There are many different definitions of "dietary fibre" in nutrition which makes comparisons somewhat confusing. Frequently the terms crude fibre, neutral or acid detergent fibre (NDF or ADF) or non-starch polysaccharides (NSP) have been used interchangeably. In biochemical terms "dietary fibre" is extremely complex and heterogeneous and in fact comprises all those polysaccharides and lignins which are not digested by the endogenous enzymes in the gastrointestinal tract. This group of materials includes, structural polysaccharides associated with plant cell wall; cellulose, pentosans, ß-glucans, and pectins. Plant cells also have structural materials which are not polysaccharides, the lignins but contribute to the dietary fibre. The cells of the gastrointestinal tract also secrete various gums and mucilages which are polysaccharide in nature but usually not digested.

These indigestible materials however may trap proteins, fats and carbohydrates and protect them from digestion in the small intestine. In particular in poultry the soluble non-starch polysaccharides (NSP) may generate increased viscosity in the small intestine and this reduces overall nutrient digestion and absorption. Use of exogenous enzymes such as cellulases, ß-glucanases and pentosanases has had significant benefits in improving nutrient utilisation in poultry.

Undigested feed components, largely dietary fibre, resistant starches, highly insoluble proteins, and potentially large quantities of absorbable nutrients often reach the large intestine where they undergo microbial fermentation. There is an enormous microbial population in the large intestine and this has to be managed for the benefit of the animal as discussed in Chapter 5. There are many different bacteria present including species of lactobacilli, streptococci, coliforms, bacteroides, clostridia and yeasts. These organisms ferment undigested feed residues into a large number of different products such as amines, ammonia, hydrogen sulphide, indole, phenol, skatole, and the volatile fatty acids, acetic, butyric and propionic.

The degree of fermentation depends primarily upon the source of dietary fibre and on the presence of nitrogen, minerals, and vitamins that are essential for the overall nutrition of the microbial population in the large intestine. The residence time of digesta is much longer in the large intestine than in the small intestine and a considerable absorption of water takes place. As the digesta moves through the gastrointestinal tract the proportion of dry matter increases. The longer residence time in the large intestine is necessary to obtain an active bacterial fermentation of dietary fibre as fibre digestion is inherently slower than is the digestion of the other non-fibrous feed ingredients.

The waste material after fermentation in the large intestine then is voided from the body as faeces. This consists of bacteria, epithelial cells and secretions of the gastrointestinal tract, inorganic salts, undigested feed residues, and water and is the final stage of the digestion process.

Absorption of digested nutrients

The nutrients, glucose, amino acids, fatty acids, vitamins and minerals, released from large feed particles by enzymatic digestion in the gastrointestinal tract are still physiologically outside the animal. They must be absorbed across the gut wall in order to be of benefit to the animal for maintenance or growth. In monogastric animals the small intestine is the main organ for absorption of nutrients. The inner surface of the small intestine is very large due to the presence of the villi. The maintenance of the villi is of major importance for good nutrient absorption. This has been repeatedly demonstrated in early weaned piglets where there is frequently atrophy of the villi during the post-weaning stress resulting in poor nutrient absorption and this can persist and reduce growth potential in later life. Residual unabsorbed nutrients are utilised for growth by pathogenic bacteria which then cause various diarrhoea syndromes.

For rapid and efficient absorption of nutrients from the lumen of the gastrointestinal tract they must first reach the wall of the small intestine and come into contact with the villi. Viscosity of the digested feed mixture is important here and particularly in poultry fed on wheat or barley diets high viscosity in the digesta can limit nutrient absorption. Supplementation of poultry diets with various enzymes, capable of reducing viscosity, have been very successful in poultry nutrition and certainly feed enzymes are an important group of nutricines for poultry.

Absorption of nutrients is a complex process involving the participation of intracellular enzymes, transport process and ion pumps. Once nutrients have reached the gut wall they are absorbed by the cells of the villi through either a passive or active transport system. Passive transport is a simple diffusion process and depends upon a concentration gradient. Passive transport requires a high concentration of nutrient in the gut lumen compared to that in the cells of the villi. Active transport requires a source of energy which is supplied as ATP, and relies upon various carriers in the membranes of the absorbing cells to transport nutrients.

Nutrients such as monosaccharides and amino acids are mainly absorbed by active transport processes. The active transport carrier in the cell membrane has two specific binding sites. The nutrient is attached to one site and the other site picks up sodium. The charged carrier moves across the intestinal membrane and deposits the nutrient and the sodium inside the cell. The empty carrier returns back across the membrane to repeat the absorptive process. Sodium which was co-transported with the nutrient is pumped out of the cell back into the gut lumen to be recycled in subsequent transport activities.

Glucose and other monosaccharides are largely absorbed by active transport. However the rates of absorption of sugars differ in decreasing order of galactose, glucose, fructose, mannose, xylose and arabinose. Some glucose that enters the epithelial cells is also utilised to provide cellular energy for maintenance of the gastrointestinal tract.

In mature animals intact dietary protein does not reach the mucosal cells of the small intestine in significant amounts. Nevertheless small amounts will escape digestion reach the mucosal cells and be absorbed. These intact proteins will then trigger immune responses and the mucosal cells will produce a range of antibodies against them (Chapter 5). Amino acids and small peptides are both absorbed by a variety of active transport carriers some of which are sodium dependent and some are sodium independent.

However, in new-born animals, nutrient absorption also occurs by pinocytosis where cells can engulf large undigested molecules, particularly proteins. This is important in new-born mammals as they obtain valuable immunoglobulins from colostrum in milk to boost their immune system.

Fatty acids and monoglycerides derived from digestion of fats and oils together with bile salts and phospholipids form micelles in the lumen of the small intestine. The micelles transport the fatty acids and monoglycerides to the wall of the small intestine where they are absorbed into the cells of the villi. After absorption these lipid materials are resynthesised into triglycerides in the form of chylomicrons which then enter the blood circulation.

Phospholipids in particular play an important role in nutrition as emulsifying agents to assist in digestion of fat. However they also help in fatty acid absorption by the formation of micelle structures. A specific class of phospholipids, the lysophospholipids are of particular interest in absorption of nutrients as they are more hydrophilic than other phospholipids. Lysophospholipids spontaneously form micelles with bile salts, fatty acids and monoglycerides. These micelles are smaller and more stable than those formed with other phospholipids found in lecithins for example.

Laboratory studies with segments or fat intestine showed that lysophosphatidylcholine enhanced absorption of oleic acid and stimulated the incorporation of fatty acids into mucosal triglycerides of rat intestine (Rampone and Long, 1977). A commercial lysophospholipid mixture has been studied in various animal species (Schwarzer and Adams, 1996) and showed useful benefits in improving nutrient utilization. Improvements in growth performance was seen with broilers and piglets. This was ascribed to the ability of lysophospholipids to improve nutrient absorption from the gastrointestinal tract that was subsequently manifested as improved growth.

Absorption of linoleic acid has an impact upon egg weight in laying hens. The source of linoleic acid for layers is dietary fat and use of a lysophospholipid mixture together with a fat blend significantly improved overall average egg mass (Bain *et al.*, 2000). Mean egg weight is certainly one of the most important factors governing the profitability of egg production and the ability to manipulate this by safe, nutritional means is a valuable tool in Total Nutrition.

Mineral absorption is another important event in the small intestine and occurs both by passive and active systems. Mineral absorption is also confounded with other biochemical events. For example calcium absorption is regulated by the active form of vitamin D, 1,25-dehydroxycholecalciferol. Low pH in the digestive tract promotes calcium absorption as does various non-digestible oligosaccharides. The presence of compounds in feed such as oxalic acid and phytic acid may reduce calcium absorption as they form highly insoluble calcium complexes. Consequently there is an important role for the enzyme phytase which hydrolyses phytic acid to release phosphorus and release calcium. The use of acid salts of calcium such as calcium formate or propionate may be helpful . These salts would be a low buffered acidic form of calcium in comparison to alkaline calcium which occurs in limestone as calcium carbonate.

Other important minerals such as zinc, iron, and copper are both essential dietary components as well as potentially toxic elements. Zinc and iron are poorly absorbed from the gastrointestinal tract and iron deficiency often leading to anaemia is a serious problem in suckling piglets. Iron transfer into the pig foetus during gestation is limited and sow's milk contains very low amounts of iron. Paradoxically piglets require substantial and increasing amounts of dietary iron due to the rapid increase in red blood cells and body mass during early growth. This problem is frequently solved by iron injections for new-born piglets but various dietary sources of iron are also important. The salt, ferrous sulphate is widely used as an iron source and feed ingredients can greatly influence iron solubility in the lumen of the gastrointestinal tract. Animals have difficulty excreting excess iron and so iron absorption is tightly regulated to prevent excess amounts being absorbed.

Mineral nutrition of animals is also an important source of environmental pollution. Large amounts of zinc oxide are used in many piglet feeds as an antibacterial agent. Copper has long been known to have growth promoting effects in pigs. However when excess levels of these minerals are fed to animals a large proportion passes through the gastrointestinal tract and is excreted in the faeces increasing environmental pollution. For this reason the EU has regulated copper content of pig feed from 35-175 mg/kg according to the age of the pigs.

A possible solution to the environmental problem is to supply the metals as organic complexes. Iron dextran is a useful iron source and other sources of metals are available such as metal propionates, amino acid complexes or protein complexes. The use of so-called "organic minerals" however is somewhat controversial in that it is difficult to demonstrate significant improvements in mineral availability when replacing inorganic metal salts with organic metal complexes (Cao *et al.*, 2000).

Malabsorption

Efficient utilization of ingested feed requires many different biochemical reactions and physiological activities. Absorption of digested feed ingredients is a crucial step in maintaining good growth rates of animals. Reduced absorption or malabsorption results in serious economic losses for the animal producer. This is particularly important in poultry where malabsorption syndrome (MAS) is a recognised problem. In this situation broilers remain alive but gain very little weight hence the colloquial term of "Runting and Stunting Disease." The aetiology of MAS is not always evident and curative programmes are difficult to implement. Piglets may suffer absorption problems due to loss of absorptive surface in the small intestine. It is well established that post-weaning stress in piglets frequently leads to reduced villus height and villus function with a subsequent impairment in the ability to absorb nutrients for rapid growth. Usually damage to the villi is never completely repaired and piglets with a reduced absorptive capacity do not catch up in their growth performance later in life.

Feed digestion and nutrient absorption in poultry

The overall digestive enzymes and biochemical processes are similar in poultry and mammals. Perhaps this is not surprising as they basically eat the same raw materials. However poultry have a different structure to the gastrointestinal tract from pigs and ruminants and this brings some specific differences in feed utilization by poultry. The digestive tract of poultry is relatively short, only 120 cm long compared to 18 m in pigs, and there is much less fermentation of feed residues in the large intestine than occurs in pigs.

The ingested feed is initially held in the crop where microbial fermentation produces acetic and lactic acids. There is a small proventriculus, equivalent to the stomach in mammals which leads into the gizzard. This has no counterpart in mammals and is concerned with mechanical breakdown of feed particles and the grinding of whole grains. Poultry have two caeca and a colon making up the large intestine but they gain little from fermentation in the large intestine. Indeed it might well be a disadvantage as gnobiotic (germ-free) birds usually grow better than normal birds. The urine and faeces are excreted together through the cloaca.

Compared with older birds the immaturity of the digestive system in young poultry results in some dietary nutrients being poorly utilized. Development of the gastrointestinal tract, nutrient transport systems, pancreatic enzyme secretion and bile salt synthesis all depend in part upon early intake of feed. This is particularly important in modern broilers where the gastrointestinal supply system must quickly mature and provide the necessary substrates for rapid muscle and bone development. Consequently early provision of nutrients affects not only immediate survival and disease resistance but also the final achievement of growth rate and bodyweights. Consequently increasing attention is now being paid to nutrition of young poultry (Dibner, 2000).

When chicks hatch, their initial source of nutrients is the portion of yolk not utilized during incubation and which may represent from 10-25% of total body weight. The yolk sac contents help the newly hatched chick adapt from the embryonic environment to independent life. Yolk sac contents are high in fat and protein and provide an important part of the chick's nutritional needs during the first few days after hatching. Yolk sac nutrients rapidly diminish and are almost depleted by the third day after hatching. Therefore in order to express their potential growth rate newly hatched chicks need to acquire the ability to digest and absorb nutrients from externally supplied feed in the first 4-5 days of life. The first five days of life has now become a large proportion of the whole life span of the modern broiler. The gastrointestinal tract must undergo very rapid growth and development in order to support the expected growth rates.

The predominant energy source for the chick changes within 2-3 days after hatching from yolk-based lipid to dietary carbohydrate and chicks are able to digest starch shortly after hatching. Various early chick feeding regimes have been tried (Vieira and Moran, 1999). Dosing of glucose by injection or oral gavage are of limited practical value and has not given clear cut benefits. Propionic acid is well known as a useful gluconeogenic substrate for ruminants and is widely used in treatments of ketosis. It is also a gluconeogenic substrate in chicks and is rapidly absorbed by the chick. Propionic acid and its salts, calcium propionate and ammonium propionate, are widely used in animal feeds as constituents of mould and bacterial inhibitors. It would seem prudent to ensure that early chick feeds are well protected against moulds and bacterial pathogens such as *Salmonella* and at the same time the propionic acid may supply energy and relieve any ketosis problems.

Young chicks have difficulty digesting and absorbing saturated fats supplied in feeds. An adequate bile supply is necessary for efficient fat digestion and absorption. It is possible that bile secretion in the duodenum of young poultry may not be adequate for efficient fat utilization, particularly if feed contains substantial amounts of highly saturated fat (Jin *et al.,* 1998). Various strategies have been tried to improve fat digestibility in chicks. Addition of bile salts increases digestibility of animal-fat blends which contain significant amounts of saturated fatty acids. Supplementation of chick feed with lipase enzymes increased fat digestibility but reduced feed intake and liveweight gain so there was no overall benefit (Al-Marzooq and Leeson, 1999). Lysophospholipids increased absorption of palmitic and stearic acids *in vitro* and this suggests they may have a beneficial role in early chick nutrition (Schwarzer and Adams, 1996).

Until the bird is about eight days of age the output of pancreatic enzymes may well limit digestion. Addition of exogenous enzymes is therefore likely to supplement the digestive capacity of the younger bird and this is an area for future research.

The yolk sac is also an important means of providing immunity until the immune system of the chick matures. Passive immunity present in the egg is dependent upon the immunological experience of the laying hen. In particular the immunoglobulin IgG provided by the hen accumulates in the yolk sac and is absorbed by the chick embryo during incubation and until two days after hatching. It is more advantageous for these antibodies to provide early immunity rather than to be used in nutrition and so it is important to supply the chick early on with a good quality digestible protein source and spare the immunoglobulins.

Some feed components, particularly the non-starch polysaccharides (NSP) and phytic acid influence digestion and absorption of nutrients in poultry. The ß-glucans in barley and arabinoxylans in wheat and rye, reduce overall nutrient digestibility in broilers (Smits and Annison, 1996). Various mannans and galacto-mannans also occur in plants and these are widely utilized in the food industry for their gel forming properties. Guar gum is a classic example. However poultry feed ingredients generally are low in mannans although a small amount is present in soyabeans. Large amounts of mannans occur in palm kernel meal and in copra but these are not widely used in poultry diets. There may be the possibility of using ß-mannanase enzymes to help deal with these non-starch polysaccharides (Jackson *et al.,* 1999). It is well established that ingestion of NSP leads to an increase in viscosity of the contents of the gastrointestinal tract. The negative effects of NSP on broiler performance is largely overcome by the use of various feed enzymes such as beta-glucanases and xylanases (Bedford, 2000). This has been so successful that enzyme addition to wheat or barley-based broiler feeds is almost routine nowadays.

One striking effect of the increased viscosity brought about by NSP is a reduction in the digestibility of fat and this is frequently improved by the addition of NSP-degrading enzymes. There is also an interaction between fat digestibility NSP-degrading enzymes and the type of fat in the diet. The ratio of unsaturated to saturated fatty acids in the diet influences digestibility of fat and metabolizable energy contents of fat or fat blends (Wiseman et al., 1991). High intestinal viscosity interferes with digestibility of saturated fats more than with digestibility of unsaturated fats. Furthermore fatty acid utlization from saturated fats such as beef tallow is improved by the addition of small amounts of unsaturated fats as found in vegetable oils. The physiological basis of this interaction between unsaturated and saturated fatty acids may be due to the greater dependence on emulsifying agents of saturated fatty acids for effective digestion and absorption.

The use of enzymes in poultry feeds in general improves the rate of nutrient utilization in the small intestine. This is important as it ensures that the products of fat, starch and protein digestion are rapidly absorbed and this gives the bird a greater competitive edge over the gut microflora, particularly as the bird matures and the gastrointestinal tract becomes more heavily colonised by bacteria. This is of even more importance when antibiotic growth promoters are no longer used as these are effective in controlling the gut microflora.

The feed mix or chyme which passes from the proventriculus and gizzard in poultry to the small intestine has a low pH and is largely devoid of competing bacteria. When the chyme enters the duodenum it is mixed with a wide range of digestive enzymes, bile acids and other secretions which also are inhibitory to rapid growth of bacteria. With highly digestible feed and a healthy well functioning small intestine feed ingredients are rapidly broken down and the nutrients absorbed for use by the bird. As the feed mix passes through the small intestine towards the large intestine there is a progressive decline in digestive enzymes and bile acid concentrations and a greater possibility for colonisation by bacteria.

With a poorly digestible feed however nutrients are not rapidly digested and absorbed by the bird and enter the lower part of the small intestine where they can be utilised by the bacteria in the gastrointestinal tract. This stimulation of bacterial inevitably allows growth of pathogenic and other undesirable bacteria. Some of these are directly associated with diseases syndromes. Proliferation of *Clostridia perfringens* for example, will lead to development of problems of necrotic enteritis and growth of *E.coli* species leads to various scouring problems. Necrotic enteritis is much more prevalent poultry fed on wheat, barley or rye-based diets than on maize-based diets which suggests that the stresses imposed by the NSP in the diet may encourage the growth of the pathogenic organism (see Chapter 5).

Antibiotic growth promoters are quite effective in controlling the growth of Gram positive organisms such as *C.perfringens* and the withdrawal of these products has made necrotic enteritis more of a threat. Other bacterial species are able to deconjugate bile acids which results in impaired fat digestion since bile acids are essential for micelle formation in the small intestine. Furthermore extensive overgrowth of bacteria in the gastrointestinal tract also requires energy and protein which reduces the amount of nutrients available to support the growth of the bird. These effects all constitute an additional stress on poultry and Total Nutrition must focus on the alleviation of stress in order to obtain good performance of poultry in the absence of antibiotic growth promoters. Application of enzyme nutricines will remain an essential part of stress reduction. Although feed enzymes are unlikely to completely solve the problems of necrotic enteritis or other enteric problems they will be an important part of the solution (Bedford, 2001).

An interesting possible explanation of the benefit of enzyme supplementation of poultry feed in controlling the gut microflora may be that the action of the enzymes on NSP and other dietary fibres produces various nondigestible oligosaccharides. As described in Chapter 5 these can be preferentially utilized by desirable ileal and caecal bacterial species which flourish at the expense of other, possibly detrimental, species. This in effect is also alleviation of stress on the birds and will promote optimal growth and health (Apajalahti and Bedford, 1999). This concept has been confirmed in rye-based broiler feeds where addition of a xylanase enzyme significantly reduced enterobacteria, total anaerobes and Gram-positive species in the gastrointestinal tract of broilers (Dänicke *et al.*, 1999).

Poultry usually consume a considerable amount of nondigestible oligosaccharides in the form of stachyose and raffinose that occur in soyabean meal and other legumes. These oligosaccharides contain galactose and are not metabolised in the small intestine due to the lack of endogenous α-1,6 galactosidase activity. They pass undigested into the large intestine and there are fermented by gas-forming bacteria. It is not clear however whether on not these galactose-containing oligosaccharides may also have modulating effects on the microflora of the gastrointestinal tract (Chapter 5).

Poultry ferment relatively small amounts of volatile fatty acids in the large intestine in comparison to pigs. Consequently these oligosaccharides may not contribute much to energy supply in poultry and this may be one reason that soyabean meals generally have a higher energy value for pigs than for poultry even though they have similar Gross Energy values (Givens and Moss, 1990) (Table 1). Comparing DE with AME is not strictly valid but as ME for pigs is about

96% of DE there is clearly significantly more energy available from soyabean meals for pigs than for poultry.

Table 1
Comparison of
energy values for
soyabean meals
determined for
pigs and for
poultry

Energy values (MJ/kg oven dry matter)	Pigs	Poultry
GE	19.3	19.7
DE	17.7	-
AME	-	11.2

There is a complex interaction however between intestinal viscosity, feed ingredients and physiological status of the bird. Intestinal viscosity of the feed mix is frequently much higher than that of the individual feed ingredients. In germ-free chicks viscosity does not develop to the same extent as in conventional birds (Langhout, 1998). Consequently benefits from enzyme addition to feeds for germ-free birds is also less than in conventional birds. This suggests that manipulation of the microflora in the gastrointestinal tract will be an important strategy in Total Nutrition, particularly in the absence of antibiotic growth promoters.

The chemical composition of the various NSPs; pentosans, ß-glucans, pectins, celluloses, and mannans are extremely diverse. These materials plus lignin, resistant starches, and poorly digestible proteins collectively make a very heterogeneous mixture that will require a wide range of different enzymes to break them down. Consequently enzymes from different origins may offer various and differing specificities and modes of action. It is unlikely however that a mixture of supplemental enzymes would be sufficiently wide ranging to completely break down all these materials to small units suitable for absorption. More likely addition of exogenous enzyme mixtures can partly break down these materials, reduce viscosity of the gut contents and possibly allow the normal digestive enzymes better access to the feed ingredients.

Phytic acid, as the phytate salt complexed with various metals, is a common ingredient in animal feeds as it occurs in all plant seeds but is poorly digested by monogastric animals. Poultry in common with other monogastric species consume a significant amount of phosphorus in the unavailable form of phytic acid. Phytase enzyme is now widely used in poultry diets (Sebastian *et al.*, 1998) and can replace added inorganic phosphorus with obvious environmental benefits. The overall economic benefit of phytase however will depend upon the relative prices of inorganic phosphorus sources and the cost of the enzyme. Certainly environmental considerations will play an increasingly important role in animal production and Total Nutrition also considers this aspect (Chapter 9).

Phytase substantially improved the amounts of phytate hydrolyzed from a wide range of feed raw materials and the total phosphorus retention in both broilers and layers (Table 2), (Leske and Coon, 1999).

Table 2

Effect of phytase
on phytate
phosphorus
hydrolysis from
raw materials and
on total
phosphorus
retention in
broilers

Raw material	Phytate hydrolysis (%)		Total phosphorus retention (%)	
	Control	Phytase	Control	Phytase
Maize	30.8	59.0	34.8	40.9
Soyabean meal	34.9	72.4	27.0	58.0
Wheat	30.7	46.8	16.0	33.8
Wheat middlings	29.1	52.2	31.9	43.4
Barley	32.2	71.3	40.3	55.5
Defatted rice bran	33.2	48.0	15.5	26.5
Canola meal	36.7	55.8	39.4	45.7

Phytase application is somewhat complicated by the fact that many feed ingredients contain endogenous phytase enzymes. Barley, rye, wheat, and triticale are quite rich in phytase whereas maize, sorghum and oilseeds are much poorer. Wheat bran can be a valuable source of phytase if it is not excessively heat treated. The location of phytic acid in various feed raw materials may also be important. Phytic acid is located in the aleurone layer in wheat, in the germ in maize and in the protein bodies in soyabeans.

Free phytic acid readily binds to proteins and this may protect the proteins from enzymatic digestion and so make them less available. Consequently there have been reports that phytase improves amino acid digestibility in poultry. An overall review of the influence of phytase on protein and amino acid digestibility indicated that addition of phytase to the feed improved digestibility of amino acids and crude protein by 1-3% (Kies et al., 2001).

The exact mechanisms whereby phytase, a phosphatase, improves nitrogen utilization is not completely established. Certainly phytate-protein complexes are present in feed raw materials, particularly soyabean meal. There is also the likelihood of de novo formation of phytate-protein complexes in the gastrointestinal tract. Phytate may also form complexes with free amino acids in the gastrointestinal tract. Proteolytic enzymes may form complexes with phytate leading to a reduction in the activity of protein digestion.

A very strong complex is formed between soluble protein and phytate at pH of 2-3 of different feed raw materials (Jongbloed et al., 1997). Pretreatment of phytate with phytase prevented the formation of this complex. After formation the hydrolysis of protein from this complex by pepsin was considerably accelerated by the action of phytase. In practice the response seems to depend upon the protein source and it is difficult to establish a solid degree of improvement to be expected in amino acid utilization from phytase addition (Bedford, 2000).

Phytic acid is a powerful chelator and forms a wide variety of insoluble salts with many different metal cations at neutral pH and this could render them unavailable and reduce absorption from the digestive tract. Zinc in particular forms a highly insoluble complex with phytic acid and this may affect zinc uptake. There is also an interaction between calcium content of the feed and phytase efficacy. High levels of dietary calcium may precipitate phytic acid as insoluble calcium phytate which may be more difficult for the phytase enzyme to hydrolyse. It is not only the absolute calcium content but its ratio to phosphorus content in the ration which is important, the greater the calcium:phosphorus ratio the poorer the response to phytase. Vitamin D also plays a role here and increased levels of this vitamin and decreased levels of calcium increases phytate utilization even in the absence of phytase addition.

Use of organic acids such as lactic or citric are effective in increasing the apparent activity of phytase. Citric acid at 4-6% of the feed was markedly efficacious in improving phytic acid utilization in broiler chicks (Boling *et al.,* 2000). Dramatic responses in weight gain and tibia ash content were observed when citric acid was added to a phosphorus-deficient maize-soyabean meal diet that contained substantial phytic acid. The increase in tibia bone ash was similar to that obtained from adding phytase enzyme to the feed. This strongly suggests that citric acid exerts a beneficial effect by increasing utilization of phytic acid. Furthermore citric acid supplementation did not elicit any response in a phosphorus-deficient but phytic acid-free diet.

The mode of action of organic acids in stimulating utilization of phytic acid is not entirely clear. They may act as chelators to keep the calcium in a more soluble form, removing it from the phytic acid which may then be more susceptible to endogenous phytase.

Feed digestion and nutrient absorption in pigs

There is an economic interest in the use of fibrous feeds for pigs. However the stomach and small intestine of pigs do not produce the endogenous enzymes required to breakdown cell wall NSP and lignins. Consequently there is a negative relationship between dietary fibre and digestibility (Lin *et al* 2000). Similar relationships were seen for dry matter, crude protein and amino acid digestibilities. The ileal apparent digestibility of gross energy decreased by 32%, while that for crude protein decreased by 12% and amino acid digestibility decreased 6%. Addition of exogenous enzymes to feeds based on barley and wheat improves both nutrient digestibility and performance of pigs, although results have not been so clear cut and consistent as with broilers.

The effect of dietary NSP on nutrient digestibility may partly be explained by an increased rate of passage through the digestive tract. Dietary fibre however also has an influence on microbial fermentation not only in the large intestine but also in the small intestine. The relatively long small intestine in the pig of 18 m compared to 120 cm for poultry increases transit time in the gastrointestinal tract. This provides increased opportunity for microbial colonisation of the small intestine.

Microbial activity in the small intestine may be assessed in terms of production of VFA. It is clear that increased microbial activity occurred in the small intestine when feeds with high levels of dietary fibre, mainly NSP from wheat, were given to growing pigs (Table 3) (Lin et al., 2000).

Table 3

Relationship between volatile fatty acid (VFA) production in the small intestine and crude fibre content in the feed

Crude fibre (g/kg)	28.9	29.0	50.2	63.8
VFA (g/kg dry matter intake)				
Acetic	2.78[a]*	2.81[a]	3.83[b]	3.82[b]
Propionic	0.51[a]	0.46[a]	1.02[b]	1.28[b]
Butyric	0.35[a]	0.30[a]	0.47[b]	0.46[b]
Total	3.64[a]	3.57[a]	5.32[b]	5.56[b]

Values in the same row with different superscripts are significantly different ($P<0.05$).

Quantitatively the total small intestine fermentation is small at about 1% of dry matter intake, when compared with microbial fermentation in the large intestine. However the increased microbial activity leads in turn to increased competition for nutrients with the host and increased levels of harmful bacterial metabolites such as ammonia, amines, bile acid derivatives and bacterial enterotoxins. All of these have a damaging effect on the mucosa of the small intestine and therefore, may reduce the rate of digestion and absorption of nutrients for the benefit of the host animal.

The strong negative relationship between dietary NSP and ileal digestibility and absorption of nutrients suggest that addition of enzymes such as xylanases to the feed could improve nutrient utilization. Xylanases could be expected to provide benefits by disruption or solubilization of cell wall polysaccharides. This would reduce or eliminate the encapsulating effects of the cell wall and possibly improve digestion and absorption of nutrients. In pigs however, unlike poultry the improvements of overall digestibility of dry matter, energy and crude protein were only of the order of 1% by addition of enzymes to the feed (Lin et al., 2000).

Pigs as well as poultry lack the 1, 6 α-galactosidase enzyme necessary to digest the α-galactosides, stachyose, raffinose which are significant components of soyabean meal. These oligosaccharides will be fermented in the large intestine into VFA. Addition of α-glactosidase to pig diets reduces VFA production in the caecum and improved overall performance suggesting that these oligosaccharides were digested by the α-galactosidase in the small intestine (Baucells *et al.,* 2000). However it is not clear whether these oligosaccharides would be of value in modulating the flora in the large intestine or whether they would be better used as an energy source directly in the small intestine. The large capacity in pigs for fermentation in the large intestine would ensure that the non-digestible oligosaccharides are utilised more so than in poultry. As discussed above (Table 1) pigs extract more energy from soyabean meal than poultry, probably due to a more efficient fermentation of these oligosaccharides into VFA.

One explanation of the difference for the results of enzyme supplementation in pigs versus poultry may be the anatomical differences between pigs and poultry. In pigs the stomach acts as the primary reservoir for ingested feed, and unlike the crop of poultry, gastric pH may decline to low levels that are detrimental to enzyme activity. The pH in the stomach of pigs fed *ad libitum* is rarely above pH 3.0. Although low pH levels may be reached in the crop and gizzard of poultry the duration of exposure to the low pH is much shorter than with pigs. Therefore a greater proportion of enzyme activity would be expected to survive in the small intestine of chickens compared with pigs. In order for exogenous enzymes to have beneficial effects they must either be protected against, or able to withstand, denaturation by low pH in the gastric region. Enzyme protection or inherent stability is therefore an important factor in determining efficacy of feed enzymes in pig nutrition.

A general conclusion is that fibre in excess of 7-10% of feed for growing pigs will inhibit growth (Varel and Yen, 1997). There are several explanations for this effect of high fibre levels; energy will be diluted, decreased digestion and absorption of nutrients and minerals will occur, and there may be increased endogenous losses in the small intestine. On the other hand high fibre may confer some benefits such as reduced gastric ulceration and less development of pathogenic bacteria that together provide useful health benefits, particularly in the absence of antibiotics.

Mature pigs and sows also have a much greater capacity to make good progress on high fibre diets than piglets. There is certainly a good potential for the creative use of novel high fibre feed ingredients in feeds for adult and reproductive animals. Some fibrous feed ingredients are readily fermented such as sugar beet pulp, soybean hulls and wheat bran and these are likely to be useful feed ingredients.

Feeding high-fibre diets to sows during pregnancy also has welfare benefits in assuaging hunger and preventing abnormal behaviour. An additional benefit here is that less stereotypic (repetitious) behaviour in sows may decrease energy requirements for maintenance and leave more energy available to support sow and piglet growth. A high-fibre diet may also produce more VFA, especially acetic acid which may be directly incorporated into milk fat and provide a higher-energy milk.

Phosphorus metabolism, phytase enzymes and utilization of phytate have been thoroughly investigated in pigs as well as in poultry. Phytase enzymes from both yeast and *Aspergillus niger* increased bioavailability of phosphorus in the diets of growing pigs (Matsui *et al.*, 200). However citric acid supplementation of feeds showed no benefit in terms of improved phytic acid utilization in pigs whilst a very positive effect was seen in poultry (Boling *et al.*, 2000).

Feed digestion and nutrient absorption in ruminants

Ruminant digestion is considerably more complex than that of monogastric animals. Young suckling ruminants have essentially a monogastric-type of digestion but as the lamb or calf begins to eat solid food the typical four-compartment rumen stomach develops.

In ruminants ingested feed is mixed with large volumes of saliva initially during eating and again during rumination. Feed is broken down partly by physical action due to contraction of the rumen walls and by the process of rumination where coarse feed materials are thoroughly chewed before being returned to the rumen.

Feed is also broken down by enzymatic digestion but now the digestive enzymes are not produced by the animal itself as in monogastrics but by the rumen microflora population. The rumen system is in effect a continuous culture system for a wide range of anaerobic bacteria and protozoa. Digested feed ingredients are fermented by the rumen micro-organisms primarily into volatile fatty acids and the gases methane and carbon dioxide. The gases are lost by eructation (belching) and the volatile fatty acids are absorbed through the rumen wall and constitute the major source of energy for ruminant animals. Some microbial cells and undegraded feed ingredients pass out of the complex stomach into the small intestine. Here they are digested by enzymes secreted by the animal itself and the digested nutrients absorbed across the wall of the small intestine.

There has been considerable interest in supplementing ruminant rations with enzymes in an attempt to improve overall feed utilization. Enzymes added to dairy cow feeds increased dry matter intake on average by 1.6 kg/day and increased milk production by 1.3 kg/day (Bauchemin and Rode, 2001). Feed enzymes can be used effectively in Total Mixed Rations (TMR) but they should be applied either to the grain or dry forage. Applying the enzyme to dry feed components creates a stable enzyme-fed complex and the enzymes will remain active in the rumen.

The mode of action of feed enzymes in ruminants is not fully established. The complete digestion of the complex mixture of dietary fibres in a normal ruminant feed would require an enormous number of different enzymes. No single feed enzyme product is likely to be able to supply the full range of enzymes needed for complete fibre digestion. Presumably supplementary feed enzymes will assist in the digestion of dietary fibre and improve the total hydrolytic capacity of the rumen as a result of synergistic effects with the rumen microbial population.

Ruminants also have a substantial capacity for fermentation in the large intestine. This can be advantageous when feed ingredients are not fully digested in the rumen and reach the caecum. The horse has an enlarged colon which is utilised for the digestion of the forage consumed.

Future directions

Considerable research energy has been devoted to devising strategies for the manipulation of rumen fermentation, and this will undoubtedly continue. In monogastrics the gastrointestinal tract is a major consumer of dietary energy. More effective modulation of the growth and development of the gastrointestinal tract will be an important route to improve animal performance in the future. Modulation of the gastrointestinal tract is not only valuable in terms of digestion and absorption of nutrients but is also extremely important in controlling various enteric diseases. Antibiotic growth promoters have played a valuable role in both improving feed utilization and modulating the flora of the gastrointestinal tract. Total Nutrition will have to devise strategies based around use nutricines and novel feed ingredients which allow high levels of animal production in an economic manner.

Novel feed ingredients

Modern feed manufacture and intensive animal production relies on a rather narrow range of raw materials for diet formulations based largely on cereals, and oilseeds. This is particularly relevant to the concerns often expressed about feeding cereals to animals rather than directly to humans the so-called "wheat or meat strategy" (Chapter 9). A wide range of fibrous feed raw materials are available that are unacceptable for direct human consumption including a variety of by-products from human food manufacture such as wheat and rice milling by-products. Many other materials are available and may be used in monogastric animal nutrition including oilseed meals, sugar beet pulp, cottonseed meal, copra meal, palm kernel meal, and a large array of forages and molasses.

Adult and reproductive pigs have a much greater ability to utilize dietary fibre than do humans because of the composition of the microflora in the large intestine. Pigs have all the major fibre degrading bacteria that are found in ruminants. Consequently pigs in the future could be raised on diets completely unsuitable for direct human consumption. The nutritional characteristics of these raw materials must be established and feed manufacturing systems developed to make use of these fibrous materials.

Poultry with a much shorter digestive tract than pigs have inherently less ability to digest fibrous materials. However raw materials of cereal origin but inedible directly for humans could possibly be processed by various enzyme treatments to improve their digestibility before being used in diet formulation for poultry.

A major innovation in terms of digestion and absorption of feed nutrients will also come from genetic engineering. Classical plant breeding has always sought to improve nutritional quality of crops. High lysine and high oil-maize lines have already been developed. Rapeseed with low erucic acid levels is a commercial success. Crop plants used as feed raw materials however still contain many antinutritional factors such as trypsin inhbitor in soyabeans, glucosinolates in rapeseed, pectins in sunflower seeds, ß-glucans in barley, pentosans in wheat and rye.

Feed raw materials of plant origin are generally low in the essential amino acids methionine and lysine. Maize for example is low in lysine whereas soyabean is low in methionine. There are many qualitative characteristics of crop plants which could be improved by genetic modification to produce feed raw materials with enhanced nutritional qualities. This has already been achieved to some extent with the development of high-lysine, high-oil maize (O'Quinn et al., 2000) and "Golden Rice" which has a high ß-carotene level. This rice will be of immense value in human nutrition since ß-carotene is a precursor of vitamin A which is frequently deficient in the diets of rice-eating people. Vitamin A is important in development and maintenance of vision in humans and clearly basic foodstuffs with an enhanced level of this vitamin will be of great value.

Minerals

Excessive use of various minerals in animal feeds is a serious source of environmental pollution and for this reason copper levels in the EU have been reduced. There is considerable interest in the use of various organic mineral complexes. In principle they offer the intriguing possibility that mineral levels could be reduced in feeds and this would certainly have a beneficial effect upon the environment.

Interactions between phytic acid, organic acids and phytase enzymes will also be worthy of further study. Organic acids are already widely used in feed hygiene programmes (Chapter 2) and in modulating the microflora of the gastrointestinal tract (Chapter 5). It is intriguing to consider an extra benefit of compounds such as organic acids that are safe, cheap and widely used directly in human food products. Organic acids will be important feed ingredients in the Total Nutrition concept.

Enzymes

Enzymes have a useful role to play as processing aids in preparation of feed ingredients. There are many different proteolytic enzymes available in commercial quantities and these have been used to digest proteins from abattoir by-products such as pig blood, poultry material and hydrolysed feathers. This enzymatically hydrolysed protein was a valuable feed ingredient for piglets and gave growth performance equal to that obtained with fishmeal (Lindemann *et al.*, 2000). This approach has additional value in that when by-products such as these are converted into a high value feed ingredient they become an economic asset rather than a source of environmental pollution.

Liquid feeding is already widely practised in pig feeding and there will be many opportunities for application of enzymes as processing aids here. Both NSP and various proteins could be pre-digested by various enzymes now commercially available.

Improved application of enzymes in feed will depend upon a better knowledge of the various substrates in feed ingredients. The susceptibility of pectins and cell wall constituents in feed raw materials to enzyme attack is still far from clear. Interactions between resistant starches, poorly digestible proteins and phytic acid conspire to make enzyme application extremely complicated. Interactions between phytic acid and phytase may have implications for the utilization of proteins by monogastric animals (Selle, *et al.*, 2000). Phytic acid combines with proteins and this may have an adverse effect upon protein digestion as well as being an unavailable source of phosphorus. Consequently the nutritional relevance of phytic acid-protein interactions will need to be considered in designing feeds which give maximum utilization of both proteins and phosphorus.

The diversity of chemical bonds in the various molecules which constitute the nutrients will require an enormous spectrum of enzymes to break them down to small units suitable for absorption from the small intestine. This is unlikely to be achieved with mixtures of commercial enzymes and so there is tremendous scope to broaden the types of enzymes which could be incorporated into feeds. This may be economically difficult to justify however, as many of the commonly used feed enzymes were initially developed for other industries such as paper making, brewing, textile processing and food manufacture. It will be interesting to see what success there will be in future developments of enzymes specifically for animal nutrition.

The characteristics of the enzymes themselves are also important. Many enzymes cannot withstand the rigours of high temperature pelleting, steam conditioning or the extrusion processes widely used in the modern feed manufacturing industry. The current solution is to use post-pelleting application of liquid enzymes. This however has the disadvantage of requiring sophisticated and expensive liquid application systems. The development of much more heat stable enzymes will allow powder formulations to be used in high temperature feed processing without the requirement of liquid application post-pelleting. Thermostable enzymes exist in nature in micro-organisms which inhabit hot springs, so-called "extremophiles" and these may well be produced in industrial scale quantities in the future.

Another strategy for improving digestibility of feed raw materials is genetic modification of crops to include digestive enzymes. Cereals or oilseeds genetically modified to produce high levels of phytase is an obvious candidate. Introduction of NSP-degrading enzymes such as ß-glucanase and xylanase into cereals might reduce the problems of high intestinal viscosity. Phospholipase A_2 converts lecithin to lysolecithin, which assists in absorption of nutrients. Feed raw material enriched in this enzyme might be better absorbed.

End products of feed digestion

Two important processes take place in the gastrointestinal tract, digestion of feed ingredients and absorption of nutrients. Digestion by its very nature however generates a whole spectrum of new molecular entities such as various peptides, oligosaccharides, derivatives of triglycerides and phospholipids. These certainly have various physiological functions as well as being sources of nutrients for the host animal and increasing emphasis will be directed to the end products of enzyme activity in the gastrointestinal tract.

Furthermore because of the molecular complexity of feed ingredients it is unlikely that supplemental feed enzymes will break down usually indigestible feed ingredients to simple monosaccharides or amino acids ready for absorption. Instead they will generate various non-digestible oligosaccharides and peptides which may be new in physiological terms to the gastrointestinal tract. Greater understanding of the role these oligomeric products play in maintaining a favourable gut microflora will form an important part of Total Nutrition in the future.

Traditionally feed proteins reaching the intestines have been regarded as sources of essential and non-essential amino acids. However given the quantity and variety of proteins consumed by animals there is enormous scope for the production of a wide range of different peptides. These peptides are often resistant to further proteolytic digestion in the gastrointestinal tract. This means that a large population of different peptides are likely to be continuously present in the gut contents. These peptides derived from feed proteins are known as bioactive peptides (BAPs) (Froetschel, 1996). The BAP can be absorbed intact and modulate digestion, appetite, immune function and endocrine metabolism by binding to specific receptors. Both plant and animal protein sources contain BAPs. Milk, wheat and soyabean proteins have been intensively studied.

One particular peptide sequence in ß-casein has received considerable research attention. This BAP was isolated from ß-casein hydrolysates and is referred to as ß-casomorphins. Released into the small intestine during the digestion of ß-casein the BAPs are absorbed intact and inhibit gastrointestinal motility and the emptying rate of the stomach.

BAPs are present in a variety of protein sources such as wheat, milk and soyabean and can be as small as tripeptides. They affect gastrointestinal motility and passage rates of digesta in different species. However the extent of their importance in nutrient-mediated control of digestive function, metabolism, and intake is unknown. The size, structure and function of these peptides from milk and other protein sources suggest that they play an ubiquitous, but in many cases an undefined role.

There is certainly the potential for greater physiological value to be obtained from dietary protein sources. Perhaps the use of specific proteolytic enzymes could generate BAPs in feeds and exert beneficial effects upon animal growth and performance. This would require considerable work to establish the presence of BAP sequences in known feed protein sources and to have available suitable enzymes to produce these BAPs in the gastrointestinal tract. This may be available in the future.

The significance of generating novel oligosaccharides from the action of NSP enzymes is completely unknown at present. The NSP-degrading enzymes have been considered beneficial because they reduce viscosity of gut contents. However they must also generate a population of novel oligosaccharides. As previously mentioned the relatively few NSP enzymes utilized will not degrade ß-glucans, pentosans, mannans, pectins or cellulose to their constituent monosaccharides but will rather generate a population of oligosaccharides and the physiological function of these is unknown.

The ability of various known non-digestible oligosaccharides to modulate the microflora in the large intestine has been extensively studied as discussed in Chapter 5. The difficulty with the products of NSP-degrading enzymes is that in most cases the end products have not been characterised and consequently any physiological response cannot be determined.

Products from the digestion of fats, oils and phospholipids also have potential to exert physiological effects. For example palm kernel oil and coconut oil are rich in the medium chain fatty acid, lauric acid and this fatty acid would be released into the gastrointestinal tract by lipase activity. Lauric acid and lauryl monoglyceride have antimicrobial activity against Gram positive bacteria (Kabara, 1978). A mixture of lauric acid, lauryl monoglyceride, caprylic acid and lactic acid was an effective antibacterial in controlling mastitis problems in dairy cattle (Bodie and Nickerson, 1992). There could well be beneficial applications of various fatty acids, particularly when antibiotic growth promoters are no longer included in the diet. Use of specific dietary oils such as palm kernel or coconut oil together with appropriate lipases might allow better control of the gut microflora to be achieved in a completely nutritional manner.

Similarly the digestion products of phospholipids could have important physiological effects. Lecithins are frequently incorporated into diets either as a source of nutrients or as emulsifiers. The action of the enzyme phospholipase A_2 on lecithins would be to produce lysolecithins which as discussed above have useful benefits as an absorption enhancer.

Total Nutrition needs to consider the end products of digestion as a source of nutrients and as a source of compounds with distinct physiological effects. It is conceivable in the future that a judicious selection of feed ingredients and of appropriate digestive enzymes will extend the value of feeds. Careful feed formulation will help the animal respond to health challenges as well as ensure that feed ingredients are efficiently digested into nutrients which are readily absorbed by the host animal.

References

Al-Marzooq, W. and Leeson, S. (1999). Evaluation of dietary

supplements of lipase, detergent,and crude porcine pancreas on fat utilization by young broiler chicks. *Poultry Science*, **78**, 1561-1566.

Apajalahti, J. and Bedford, M. (1999). Improve bird performance by feeding its microflora. *World Poultry*, **15**, 20-23.

Bain, W. M., Wakeman, W. and Solomon, S. E. (2000). The effects of the inclusion of phosphatidyl choline on egg production and quality. *World Poultry Science Association Meeting*, Montreal, Canada.

Baucells, F., Perez, J. F., Morales, J. and Gasa, J. (2000). Effect of α-galactosidase supplementation of cereal-soya-bean-pea diets on the productive performances, digestibility and lower gut fermentation in growing and finishing pigs. *Animal Science*, **71**, 157-164.

Beauchemin, K. A. and Rode, L. (2001). Feed enzymes for ruminants-from research to field. *Feed Mix*, **9**, 30-31.

Bedford, M.R. (2000). Exogenous enzymes in monogastric nutrition – their current value and future benefits. *Animal Feed Science and Technology*, **86**, 1-13.

Bedford, M. (2001). Enzymes, antibiotics and the intestinal microflora. *Feed Mix*, **9**,32-43.

Boddie, R. L. and Nickerson, S. C. (1992). Evaluation of postmilking teat germicides containing Lauricidin, saturated fatty acids and lactic acid. *Journal of Dairy Science*, **75**, 1725-1730.

Boling, S. D., Webel, D. M., Mavromichalis, I., Parsons, C. M. and Baker, D. H. (2000). The effects of citric acid on phytate-phosphorus utilization in young chicks and pigs. *Journal of Animal Science*, **78**, 682-689.

Cao, J., Henry, P. R., Guo, R., Holwerda, R. A., Toth, J. P., Littell, R. C., Miles, R. D. and Ammerman, C. B. (2000). Chemical characteristics and relative bioavailability of supplemental organic zinc sources for poultry and ruminants. *Journal of Animal Science*, **78**, 2039-2054.

Dänicke, S., Vahjen, W., Simon, O., and Jeroch, H. (1999). Effects of dietary fat type and xylanase supplementation to rye-based broiler diets on selected bacterial groups adhering to the intestinal epithelium, on transit time of feed and on nutrient digestibility. *Poultry Science*, **78**, 1292-1299.

Dibner, J. (2000). Early nutrition in young poultry. In: *Recent Advances in Animal Nutrition*. Eds: P. C. Garnsworthy and J. Wiseman. Nottingham University Press, UK. pp. 73-88.

Froetschel, M, A. (1996). Bioactive peptides in digesta that regulate gastrointestinal function and intake. *Journal of Animal Science*, **74**, 2500-2508.

Givens, D. I. and Moss, A. R. (1990). *UK Tables of Nutritive Value and Chemical Composition of Feedstuffs*. Rowett Research

Services Ltd., UK.

Jackson, M. E., Fodge, D. W. and Hsiao, H. Y. (1999). Effects of beta-mannanase in corn-soybean meal diets on laying hen performance. *Poultry Science*, **78**, 1737-1741.

Jin, S-H., Corless, A. and Sell, J. L. (1998). Digestive system development in post-hatch poultry. *World's Poultry Science Journal*, **54**, 335-345.

Jongbloed, A. W., De Jonge, L. H., Kemme, P. A., Mroz, Z. and Kies, A. K. (1997). Non-mineral effects of phytase in pig diets. In: *Proceedings of the 6th Forum on Animal Nutrition*, BASF, Germany. pp. 92-106.

Kabara, J. J. (1978). Fatty acids and derivatives as antimicrobial agents-A review. In: *Symposium on the Pharmacological Effect of Lipids*. Ed: J. J. Kabara. The American Oil Chemist's Society, USA.

Kies, A. K., Van Hemert, K. H. F. and Sauer, W. C. (2001). Effect of phytase on protein and amino acid digestibility and energy utilization. *World's Poultry Science Journal*, **57**, 109-126.

Langhout D. J. (1198). *The role of the intestinal flora as affected by non-starch polysaccharides in broiler chickens*. PhD Thesis. Wageningen Agricultural University, Wageningen, The Netherlands.

Leske, K. L. and Coon, C. N. (1999). A bioassay to determine the effect of phytase on phytate phosphorus hydrolysis and total phosphorus retention of feed ingredients as determined with broilers and laying hens. *Poultry Science*, **78**, 1151-1157.

Lindemann, M. D., Cromwell, G. L., Monegue, H. J., Cook, H., Soltwedel, K. T., Thomas, S. and Easter, R. A. (2000). Feeding value of an enzymatically digested protein for early-weaned pigs. *Journal of Animal Science*, **78**, 318-327.

Matsui, T., Nakagawa, Y., Tamura, A., Watanabe, C., Fujita, K., Nakajima, T. and Yano, H. (2000). Efficacy of yeast phytase in improving phosphorus bioavailability in a corn-soybean meal-based diet for growing pigs. *Journal of Animal Science*, **78**, 94-99.

O'Quinn, P. R., Nelssen, J. J., Goodband, R. D., Knabe, D. A., Woodworth, J. C., Tokach, M.D. and Lohrmann, T. T. (2000). Nutirional value of a genetically improved high-lysine, high-oil corn for young pigs. *Journal of Animal Science*, **78**, 2144-2149.

Rampone, A. J. and Long, L. R. (1977). The effect of phosphatidylcholine and lysophosphatidylcholine on the absorption and mucosal metabolism of oleic acid and cholesterol in vitro. *Biochimica and Biophysica Acta*, **486**, 500-510.

Schwarzer, K and Adams, C. A. (1996). The influence of specific phospholipids as absorption enhancer in animal nutrition. *Fett/Lipid*, **98**, 304-308.

Sebastian, S., Touchburn, S. P., and Chavez, E. R. (1998). Implications of phytic acid and supplemental microbial phytase in poultry nutrition: a review. *World's Poultry Science Journal*, **54**, 27-47.

Selle, P. H., Ravindram, V., Caldwell, R. A. and Bryden, W. L. (2000). Phytate and phytase: consequences for protein utilization. *Nutrition Research Reviews*, **13**, 255-278.

Smits, C.H.M. and Annison G. (1996). Non-starch plant polysaccharides in broiler nutrition – towards a physiologically valid approach to their determination. *World's Poultry Science Journal*, **52**, 203-221.

Varel, V. H. and Yen, J. T. (1997). Microbial perspective on fibre utilization by swine. *Journal of Animal Science*, **75**, 2715-2722.

Vieira, S. L. and Moran Jr., E. T. (1999). Effects of egg of origin and chick post-hatch nutrition on broiler live performance and meat yields. *World's Poultry Science Journal*, **55**, 125-142.

Wiseman, J., Salvador, F. and Craignon, J. (1991). Prediction of the apparent metabolisable energy content of fats fed to broiler chickens. *Poultry Science*, **70**, 1527-1533.

Yin, Y-l., McEvoy, J. D. G., Schulze, H., Hennig, U., Souffrant, W-B., Mccracken, K. J. (2000). Apparent digestibility (ileal and overall) of nutrients and endogenous nitrogen losses in growing pigs fed wheat (var. Soissons) or its by-products without or with xylanase supplementation. *Livestock Production Science*, **62**, 119-132.

5   Struggle for Supremacy: Management of the Gastrointestinal Tract

The gastrointestinal tract is the largest organ in the body and a major consumer of metabolic energy, using about 20% of all ingested energy (Cant et al., 1996). In monogastrics it has the highest rate of protein synthesis of all the body tissues and synthesises more than 20% of the daily whole body proteins (Van Der Meulen and Jansman, 1997). It is the largest endocrine gland in the body and produces at least 20 hormones, regulatory peptides and their receptors. It is the largest site of the immune system in the body and is the location for the majority of the lymphocytes and other immune cells in the gut-associated lymphoid tissue (GALT).

The gastrointestinal tract is also a curious organ in that two different structures and modes of operation have evolved as seen in ruminant and monogastric animals. In all species two major biochemical processes occur in the gastrointestinal tract, enzymatic digestion (see Chapter 4) and microbial fermentation. In ruminants fermentation takes place in the rumen before enzymatic digestion in the small and large intestines. In monogastric animals the fermentation processes occur in the large intestine after enzymatic digestion of feed in the stomach and small intestine. In ruminants fermentation of feed is of major importance in supplying nutrients, whereas in monogastric species fermentation plays a minor role in releasing nutrients for the host animal (Ewing and Cole, 1994).

The gastrointestinal tract harbours an enormous microbial community with the number of microbial cells vastly greater than the total cells of the body of the host animal. In humans, over 90% of the total cells in the entire body are present as bacteria in the colon (Gibson and McCartney, 1998). In a study on the microflora in the caecum and faeces of rats 627 colonies were identified at the species level (Kleessen et al., 1997).

This microbial population or ecosystem as it is sometimes described may not be essential for the survival and growth of animals and indeed germ-free (gnobiotic) chickens grow better than conventional birds, even those treated with antibiotic growth promoters. Germ-free animals have several different characteristics in the gastrointestinal tract compared to conventional animals (Ewing and Cole, 1994). The total weight of the small intestine is lower in germ-free animals with a lower metabolic energy cost and a decrease in endogenous losses. In general it seems that any benefits obtained from the microbial population present in the gastrointestinal tract of monogastric animals is outweighed by the disadvantages of metabolic cost and the risk of enteric diseases. This is not true in ruminants where the rumen microflora permit the ruminant animal to survive on poor quality high cellulose forage materials.

This microbial population in the gastrointestinal tract of monogastric animals frequently acts in an autonomous manner that may be deleterious to the well-being of the animal. In some respects the microbial population of the gastrointestinal tract can be viewed as another organism often in competition with the host animal. It is an interesting question as to whether the gastrointestinal tract of an animal is a integral part of the animal or whether the animal is merely a structure to shelter and feed the microflora in the gastrointestinal tract.

One of the major challenges in Total Nutrition is to manage this "competing organism" for the benefit of the host. The gastrointestinal tract of any animal is a large organ and is expensive to maintain in terms of energy and protein requirements. It uses an enormous amount of nutrients for maintenance, tissue renewal and processing of nutrients. Any improvements to the health and efficiency of this organ would certainly lead to an overall improvement in the health and the productivity of the animal. Indeed various enteric diseases are a major scourge of all animal species including humans.

The stability of microbial fermentation in the gastrointestinal tract of monogastric animals influences the establishment of bacterial pathogens since a stable intestinal fermentation is resistant to the establishment of pathogens (Hillman *et al.*, 1994). Fermentative stability is difficult to measure in animals as changes within the gastrointestinal contents over time would need to be studied. However, certain groups of micro-organisms in the gastrointestinal microflora, in particular the lactic acid bacteria are known to be effective in the inhibition of pathogens. Consequently an estimate of the potential efficacy of the population in the gastrointestinal tract to resist pathogen could be based on the proportions of lactobacilli:coliform bacteria in the gut contents. In general a relatively large population of lactobacilli in relation to coliform bacteria would indicate increased numbers of those bacteria inhibitory to pathogens and this may be a useful indicator of animal health.

Virus infections of the gastrointestinal tract are also a major challenge to the health of animals. Both pigs and poultry are susceptible to a wide range of enteric viral diseases. The outcome of these infections is determined by a variety of factors such as age of animal, immune status, nutrition and environment. In addition virus infections of the gastrointestinal tract are likely to lead to the development of other disease syndromes. Virus-induced damage to the mucosa of the gastrointestinal tract may provide a port of entry for other pathogens such as *E. coli* or

Salmonella. Viral damage to the gastrointestinal tract will also lead to poor digestion of feeds and malabsorption of nutrients resulting in nutritional deficiencies. This will have a cascade effect of reducing health, growth and the resistance of the animal to other infections.

The influence of dietary changes on intestinal epithelial differentiation and growth are especially marked at birth and weaning. The early weeks of neonatal life see extensive changes in gut morphology, increases in *de novo* protein synthesis and both age-related and diet induced changes in digestive functions. During the same period the rapidly changing mucosal surfaces become colonised by successions of gut bacterial groups. In the majority of animals the dynamic balance between host physiology, diet and gastrointestinal microflora leads to the establishment of a stable microbial ecology characterised by the presence of commensal organisms which exert a positive influence in maintaining and establishing a healthy gut immune system. However perturbation of the gut ecosystem can often occur in neonates and the pre-weaning period still represents the time of greatest morbidity and mortality. Poor nutrition can lead to enteric infections, resulting in mucosal damage and scours in piglets and other young animals. These disease symptoms are exacerbated by the relative immaturity of the developing mucosal immune system and the virulence of opportunistic pathogens that colonise the rapidly changing mucosal surface in the young animal.

The gastrointestinal tract is a large consumer of energy for its own metabolism and therefore shifts in gut function may affect animal performance. In the small intestine some microbiological fermentation takes place and the micro-organisms also compete with the host animal for readily digestible nutrients and as much as 6% of the net energy in a pig feed can be lost due to microbial fermentation in the small intestine (Mikkelsen and Jensen, 2000). Manipulating small intestinal transport rates and transport efficiency present opportunities to improve livestock performance. In animals consuming energy-dense diets augmentation of transport capacity of glucose and other nutrients may reduce the quantity of glucose that flows into the large bowel where it is inefficiently processed.

The overall picture is illustrated in Figure 1. Pathogenic micro-organisms ingested with food and water cause initial damage to the mucosa of the small intestine. The injured mucosa then leads to a reduction in growth from at least two events. Firstly atrophy of the intestinal villi results in a loss of enzymes that are required for normal digestion and absorption of nutrients. Secondly, the mucosal damage interferes with the barrier function of the small intestine, allowing the uptake of antigenic macromolecules into both the mucosal tissues and to the blood. These absorbed materials stimulate the immune and inflammatory responses leading to further damage of the intestinal mucosa.

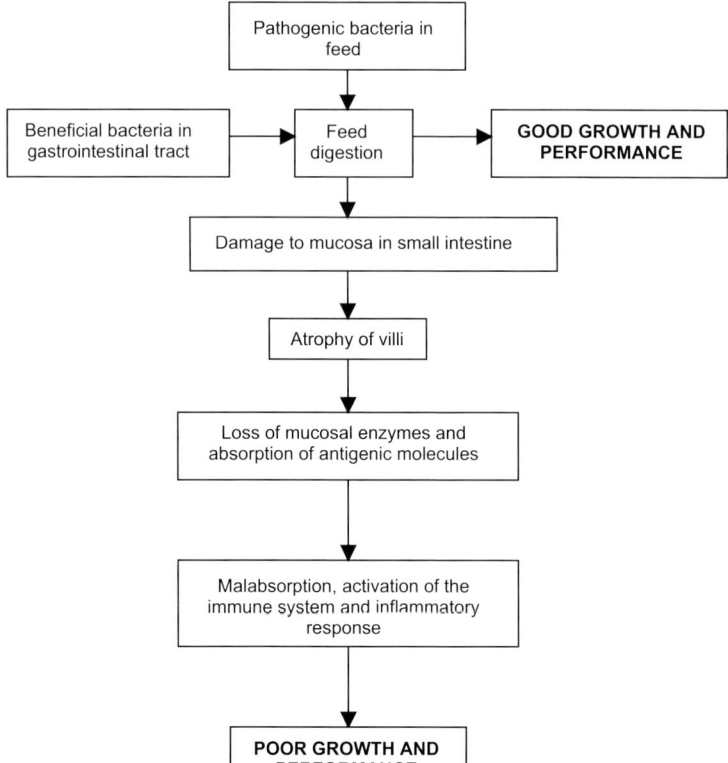

Figure 1
Relationship between feed-borne pathogens, beneficial microflora and animal growth and performance

The gastrointestinal microflora, digestive physiology, immune stimulation, and inflammation are all more or less influenced by feed ingredients and nutrition in general. The final result is either good health and growth or disease and poor productivity. The challenge of Total Nutrition is to ensure that good health and growth prevails.

The major objective of Total Nutrition in managing the gastrointestinal tract is to establish conditions which favour the growth of beneficial bacterial species such as *Lactobacilli* and *Bifidobacteria* which in turn will inhibit the growth of pathogenic organisms. It is also important to reduce stress on the animals since diseases such as necrotic enteritis in broilers or scouring in piglets are exacerbated by stress. This stress may arise from feeding cereals such as wheat or barley that are difficult to digest. Various strategies are available to help an animal resistant pathogens without recourse to antibiotics or other medication.

Protein degradation in the large intestine should be reduced and adequate glutamine supplied as an important energy source for the cells of the gastrointestinal tract. It is helpful to reduce the supply of readily available carbohydrates to the large intestine but to supply fermentable carbohydrates (prebiotics) to encourage beneficial bacteria. Organic acids may be incorporated into feeds that are inhibitory to pathogenic micro-organisms. Desirable bacteria can be supplied in feed (probiotics) to help the animal resist the pathogenic organisms. Digestive stress can be alleviated by incorporation of enzymes into feeds as discussed in chapter 4.

Promoting animal health and improving growth rate and feed efficiency through modification of the microbial fermentation in the gastrointestinal tract of food animals is a topic of major importance. This is greatly influenced by proteins and carbohydrates in feed composition. Several nutricines such as enzymes, phospholipids, and organic acids also play an important role in managing the gastrointestinal tract.

The growth, development and intrinsic differentiation of the digestive tract in neonatal pigs is profoundly influenced through interaction with dietary constituents and the microflora. For example colostrum and milk contain high levels of growth factors that accelerate proliferation and maturation of the gut in neonatal animals. Commensal bacteria play a beneficial role in modulating intestinal epithelial differentiation and promoting immune competence. For the pre- and post-weaning piglet the nature of these interactions and their impact upon enteric development and health is important in the design of novel strategies to enhance natural disease resistance and growth performance.

Damage to the mucosal barrier of the gastrointestinal tract allows translocation of macromolecules into the mucosa and blood, triggering both local and systemic immune and inflammatory mechanisms. Partial villus atrophy reduces absorption and digestion of lactose and other nutrients.

A healthy gastrointestinal tract is based upon a good diet and this in turn leads to good animal health. In Total Nutrition where animal health has to be maintained through feeding it will be important to design feeds that inhibit the proliferation of pathogenic bacteria in the gastrointestinal tract. This can be achieved by considering protein and carbohydrate metabolism on the gastrointestinal tract by feeding organic acids as inhibitors of pathogenic bacteria or by the use of competing bacteria as probiotics.

Protein and amine metabolism

Protein metabolism in the gastrointestinal tract is very active. The high protein synthesis rate of the gastrointestinal tract is largely due to the synthesis of endogenous proteins. These are proteins synthesised by the animal and secreted into the lumen of the gastrointestinal tract. These include proteins in gastric and intestinal juice and in the mucosa. The process of synthesis and secretion of endogenous proteins is largely influenced by feeding level and diet composition. Anti-nutritional factors such as protease inhibitors, lectins, tannins and non-starch polysaccharides increase the synthesis and secretion of endogenous proteins in the gastrointestinal tract. These endogenous proteins however are not necessarily lost as a significant amount, probably 70-79% is re-absorbed in the small intestine (Van Der Meulen and Jansman, 1997).

Protein metabolism is also important in the large intestine where the resident bacterial population breaks down proteins for energy whenever the concentration of carbohydrate is insufficient. This may result in the production of toxic materials such as indoles, amines, ammonia and phenols. However if an adequate level of carbohydrate can be maintained in the large intestine it could stimulate the *Lactobacilli* that in turn could inhibit the pathogens such as coliforms and the negative effects of protein breakdown could be avoided. Feeding piglets on retrograded waxy maize starch which is poorly digested in the small intestine increased the *Lactobacilli:Coliform* ratio very significantly (Reid and Hillman, 1999). This indicates that it is possible to manipulate the microflora in the gastrointestinal tract in order to have better pathogen resistance.

The presence of fermentation substrates in the large intestine may not always be desirable however as swine dysentry caused by the organisms *Serpulina hyodysenteriae* may be exacerbated by the presence of fementable substrates, particularly NSP but with the possible involvement of resistant starch (Pluske *et al.*, 1996).

Amines

Feed raw materials such as cereals contain relatively large amounts of various polyamines and therefore they are a common constituent of feeds. The amines also possess biological activity (Bardócz *et al.*, 1998). They can be classified into three groups: biogenic amines (histamine, tyramine, tryptamine, serotonin, phenylethylamine, putrescine, cadaverine and agmatine), polyamines (spermine, spermidine, putrescine, cadaverine and agmatine) and glutamine. All these amines may occur in normal feeds but whilst polyamines and glutamine appear to be essential for health biogenic amines are usually detrimental.

Biogenic amines

Biogenic amines are synthesised by bacteria through decarboxylation of amino acids. Consequently all feedstuffs produced by fermentation processes or exposed to bacterial contamination during storage can contain substantial quantities of biogenic amines. However the biogenic amines putrescine, cadaverine and agmatine do not result from the action of bacterial decarboxylase but are synthesised by plants.

Biogenic amines cause a variety of effects on health and well-being of both animals and humans. High biogenic amine consumption can be toxic and lead to nausea, respiratory distress, sweating, and heart palpitations. Therefore the biogenic amine content of feeds and also of human foods should be kept as low as possible.

High protein meals are always a potential source of biogenic amines and this is frequently controlled in meat and bone meals to ensure low levels.

Polyamines

Polyamines play an important role in nearly every step of DNA, RNA and protein synthesis and therefore are essential for growth of living organisms. Polyamines are very stable compounds which resist heat and survive acidic conditions. In mammals polyamines might arise from; (1) *de novo* synthesis, (2) the feed, (3) bacteria resident in the gastrointestinal tract. However it appears that feed is a very important and perhaps major source of polyamines.

Glutamine metabolism

Glutamine is an important respiratory fuel for cells of the gastrointestinal tract although cells of the large intestine use butyric acid as a major energy source. Glutamine can be supplied both from the intestinal contents and by the arterial blood.

Glutamine is not solely an energy source as it is also important in general nitrogen metabolism and nucleotide synthesis. Glutamine is metabolized by tissues of the gastrointestinal tract into ammonia, glutamic acid, alanine, aspartic acid, ornithine, citrulline and proline. Glutamine is probably more important in young pigs as a metabolic energy source than in mature pigs.

Carbohydrates

Animals consume a wide range of different carbohydrates which can be broadly classified into two groups:

1) DIGESTIBLE CARBOHYDRATES.

These are, starch, sucrose and lactose which are readily digested by endogenous enzymes in the small intestine such as α-amylase, sucrase and lactase. These are a major energy source for monogastric animals.

2) DIETARY FIBRE

This includes various non-starch polysaccharides (NSPs), non-digestible oligosaccharides (NDOs), pectins, resistant starch and cellulose. These materials are not digested in the upper gastrointestinal tract in monogastric animals.

Non-starch polysaccharides (NSP)

The NSPs such as ß-glucans from barley and pentosans from wheat or rye and pectins which occur in many feed ingredients frequently form gels in the small intestine in monogastric animals and this interferes with digestion and nutrient absorption as discussed in Chapter 4. Perhaps the most obvious effect of NSPs in the diets of poultry is an increase in litter moisture and the excretion of sticky droppings. This has further adverse effects in terms of increased ammonia generation from the litter and subsequent problems of animal health due to increased incidences of hock burns and breast blisters in broilers. Addition of enzymes to feeds has been very successful in reducing the viscosity of these gels and improving nutrient absorption.

The ingestion of soluble NSPs by piglets does not seem to induce the same viscous-producing properties as in other monogastric animals such as broiler chickens and furthermore certain NSPs may positively influence gut health by serving as a substrate for the proliferation of beneficial bacteria in the gastrointestinal tract.

These carbohydrates which escape enzymatic digestion in the small intestine undergo anaerobic fermentation by the microflora of the large intestine. This results in the production of volatile fatty acids; acetic, propionic, and butyric acids together with lactic acid, gases such as methane, and the microbial biomass. Lactic acid is not a volatile fatty acid and so the combined organic acids produced in the large intestine are also referred to as short chain fatty acids (SCFAs). An increase in acidity inhibits the growth of pathogenic organisms which may be pH sensitive and also improves the muscular tone of the large intestine. Uptake of these SCFAs is accompanied by uptake of water and of electrolytes including calcium. It seems that uptake of calcium may be enhanced by the presence of non-digestible oligosaccharides in the feed that presumably result in increased SCFA production. The SCFAs are rapidly absorbed, providing important energy supplies for colorectal epithelium. They promote colonocyte proliferation and assist in reversing atrophy of cells of the large intestine which can be a feature of low fibre diets. Of the SCFAs, butyric acid is considered to be the preferred energy source accounting for approximately 70% of the total energy consumed in colonocytes. Butyric acid is the main source of energy for the normal colon and *in vivo* it stimulates cell proliferation, although curiously enough *in vitro* it often appears to be an inhibitor of cell division (Gálfi and Neogrády, 2001). An impaired ability of cells to metabolise butyric acid contributes to the inflammation of the colonic mucosa resulting in ulcerative colitis. It is probably important to maintain an active production of SCFAs in the large intestine and this will be determined by feed composition.

Non-digestible oligosaccharides

These are another important sub-group of feed carbohydrates that occur in many different legume seeds, in palm kernels and copra and they are also produced by fermentation reactions (Ijl and Tivey, 1998). These are extensively used in human nutrition as a component of "Functional Foods " and have a wide range of activities including modifying the microflora of the gastrointestinal tract, possibly improving calcium absorption and influencing blood lipids (Adams, 1999).

Non-digestible oligosaccharides are often described as prebiotics which indicates that they are non-digestible feed ingredients that beneficially affect the host animal by selectively stimulating the growth of desirable bacteria in the large intestine and consequently improve animal health (Gibson and Roberfroid, 1995).

Non-digestible oligosaccharides based on the sugar mannose may also offer protection by acting as ligands for pathogenic bacteria and preventing their adhesion to or invasion of the wall of the gastrointestinal tract. The concept is that pathogenic bacteria would adsorb to a mannanoligosacchride instead of attaching to intestinal epithelial cells. If bacteria cannot adhere to the cells of the gastrointestinal tract they cannot develop pathogenicity.

Results with turkeys have indicated that mannanoligosaccharides improved performance comparable to that seen with antibiotic growth promoters (Parks *et al.*, 2001). This improvement in performance may be due to a reduction in the pathogen load encountered by the birds under farm conditions although this has not been directly demonstrated and would be difficult to do. Furthermore in order to fulfil this role of pathogen binding the oligosaccharides must be resistant to digestion in the small intestine and this also needs to be established.

Resistant starch

Feeding of resistant starch is another possibility to influence microbial metabolism in the gastrointestinal tract. Resistant starch is defined as starch that escapes digestion in the small intestine and provides a source of fermentable substrate for the microflora of the large intestine. This anaerobic fermentation leads to the production of short chain fatty acids which can be used by the animal as an energy source.

Resistant starch may be starch granules that are physically inaccessible to the digestive enzymes in the small intestine. This can arise from feeding whole cereals or when there is high viscosity in the small intestine due to the presence of non-starch polysaccharides in the diet. Resistant starch also occurs in specific starch granules such as potato starch, or it may be retrograded starch generated through feed processing. However the amount of resistant starch in a diet is likely to vary widely. The amount of starch that escapes digestion in the small intestine will depend upon several factors such as transit time, physical accessibility, amylase activity, the presence of enzymes inhibitors and the interaction of starch with other dietary components such as protein, fat and fibre. Some starch is not even fermented in the large intestine and will appear in the faeces.

Feeding resistant starch to rats increased the volatile fatty acid content in the faeces (Kleessen *et al.*, 1997). As shown in Table 1, acetic acid was the predominant volatile fatty acid but with 10% resistant starch in the form of modified potato starch significant increases in propionic and butyric acids were seen. There was probably also an increase in lactic acid production although this was not measured.

Table 1
Effects of resistant starch (modified potato starch) on volatile fatty acids (micromoles/g) in rat faeces over a five month period

Volatile fatty acid	Treatment	
	Control	Resistant starch
Acetic	43.4	88.4
Propionic	4.1	16.7
Butyric	4.3	12.3

Feeding resistant starch generally increase the populations of *Bifidobacteria* and *Lactobacilli* in the large intestine and these are usually considered to be beneficial micro-organisms. These micro-organisms increase resistance to enteric disease by reducing the growth of pathogenic bacteria. This may be achieved by lowering the pH, competing directly for substrates and for mucosal attachment sites, producing inhibitory molecules such as bacteriocins and stimulating the gastrointestinal immune system. Consequently resistant starch may be a useful ingredient in animal feeds that leads to an increase in the concentration of short chain fatty acids in the large intestine and stimulates the growth of certain bacteria which promote health.

A disadvantage of feeding either resistant starch or NSPs to pigs is that these materials tend to increase large intestine weights (Pluske *et al.*,

1998). This is economically important as it usually results in decreased dressed carcass weight of the pigs. As so often in nutritional sciences this poses another paradox. Resistant starch may be useful in maintaining health of the animals but may also reduce dressed carcass weight.

Interactions of carbohydrates and pathogenic bacteria

Many enteric pathogens use carbohydrate-binding proteins to attach to cells of the gastrointestinal tract and initiate disease. These pathogen binding proteins need specific carbohydrate ligands usually consisting of various monosaccharides such as galactose, fucose and mannose (Steer et al., 2000). Soluble oligosaccharides may prevent bacterial attachment and dislodge bacteria already attached to cells of the wall of the gastrointestinal tract. Binding by pathogenic bacteria could be blocked by incorporation of oligosaccharides into the diet.

Feeding poultry relatively high levels of different sugars such as arabinose, galactose, lactose and mannose reduces the ability of Salmonella species to adhere to epithelial cells of chickens and consequently to colonize the gastrointestinal tract (Oyofo et al., 1989a; Oyofo et al., 1989b). However treatment with these sugars does not entirely eradicate Salmonella contamination and this is in practice a major objective. Another practical disadvantage with this programme is that large amounts of sugars need to be administered to birds which would be prohibitively expensive in practice.

An alternative strategy is to use mannose-based carbohydrates that occur naturally. For example yeast cell walls, palm kernel meal and copra meal contain high levels of mannanoligosaccharides. A mannanoligosacchardie from yeast was able to reduce caecal Salmonella typhimurium concentrations in chicks but not the amounts of total coliforms (Spring et al., 200). This is an encouraging result and the fact that the mannanoligosaccharide had no effect upon Lactobacilli, lactic acid, volatile fatty acids or caecal pH suggests that its effect was that of binding the pathogen and not being involved in fermentation to produce inhibiting organic acids. It may be possible to develop some cost-effective oligosaccharide preparations which could assist in controlling pathogens in the gastrointestinal tract.

Organic acids

The different non-digestible carbohydrates that enter the large intestine are fermented into various organic acids, primarily acetic, butyric, lactic and propionic. These may encourage the growth of non-pathogenic bacteria by acting as a substrate or carbon source and also inhibit the growth of pathogenic bacteria.

The pH in the caeca of broilers during their growth varies from 5.5-5.0 and is maintained at these low values by the presence of lactic, acetic, propionic and butyric acids (Van der Wielen *et al.*, 2000). Furthermore there were significant correlations between the numbers of *Enterobacteriaceae* which include *Salmonella*, and the concentrations of acids in the caeca. Growth of the *Enterobacteriaceae* was greatly inhibited by these organic acids but they had no effect upon the growth of *Lactobacilli*. This suggests it will be important in Total Nutrition to encourage adequate levels of these organic acids in the caeca as a means of reducing pathogenic bacterial loads.

Fortunately these organic acids produced in the gastrointestinal tract; acetic, butyric, propionic and lactic are not inhibitory to the *Lactobacilli*. Consequently they may shift the microbiological balance in the gastrointestinal tract in favour of the *Lactobacilli* which will also likely increase the lactic acid content in the large intestine through fermentation processes of the gastrointestinal microflora. This is likely to be beneficial since lactic acid itself is quite effective in controlling various Gram negative pathogens such as *Salmonella* and *E. coli* species. Piglets have considerable lactic acid in the gut contents during the suckling phase. This rapidly decreases after weaning (Bolduan *et al.*, 1988) and under normal feeding conditions the intestinal concentration of lactic acid continues to decreases with age and is negligible at 49 days of age (Barnes *et al.*, 1979).

It is hardly surprising therefore that nutritionists have long used various organic acid supplements in animal feeds in an attempt to manage the microflora in the gastrointestinal tract. Organic acids can ionise in the gastrointestinal tract and supply hydrogen ions over a wide range of pH to help the gastric acid supply. The ionised organic acids remain as anions such as lactate or fumarate in the gastrointestinal tract where they can exert an antimicrobial effect and ultimately be utilised as a source of metabolisable energy.

Various organic acids can be added directly to feed or produced by fermentation in a liquid feed (Geary *et al.*, 1999; Mikkelsen and Jensen, 2000). Fermented feed has high numbers of *Lactobacilli* and yeasts with low pH, around 4.0, and a high concentration of lactic acid. Such a mixture is likely to reduce scouring problems in pigs and reduce general digestive disorders and may offer useful benefits, although a disadvantage is the necessity to manage a fermentation system on the animal production site.

Whilst organic acids in feed can slightly reduce the pH of the feed, they are unlikely to dramatically shift the pH of the gastrointestinal tract, which is a highly buffered system.

This is illustrated in Table, 2, where feeds with either 1.5% fumaric or 1.5% citric acid were given to piglets (Risley *et al.*, 1991). The organic acids did not reduce the pH in either the gastric digesta or in the digesta of the other intestinal sections.

There has been a considerable amount of effort expended in studying the effect of various organic acids upon nutrient digestibility in feeds. For example, supplementation of diets for early weaned piglets with fumaric acid has shown improvements in ileal digestibility of crude protein and amino acids (Blank *et al.*, 1999). The magnitude of the response however is likely to be influenced by differences in feed ingredients, buffering capacity of the feed, age of the animals and type and level of the acids used.

Table 2

Gastrointestinal pH in eight week old piglets given either 1.5% fumaric or 1.5% citric acids in feed

Section of gastrointestinal tract	Treatment		
	Control	Fumaric acid	Citric acid
Stomach	4.73	4.30	4.83
Jejunum	7.06	7.01	7.00
Caecum	5.96	6.04	6.05
Lower colon	6.51	6.53	6.47

The main benefit of organic acids is more likely to come from controlling the microflora in the gastrointestinal tract. In particular they can reduce the coliform burden in the gastrointestinal tract and this would be a positive benefit in terms of animal health (Mathew *et al.*, 1996, Overland *et al.*, 2000). They may also have an immunomodulating effect and reduce unnecessary stimulation of the immune system by controlling pathogenic bacteria. The organic acids may help in gastric digestion of proteins particularly in piglets and improved apparent ileal digestibilities of protein and amino acids have been observed in fattening pigs. Organic acids may positively influence the mucosal structure in the gastrointestinal tract and stimulate pancreatic secretion of digestive enzymes. They may also assist in the digestion and absorption of many nutrients. Various organic acid combinations have been shown to improve the feed:gain ratio of both weaned and fattening pigs and in general organic acids are useful alternatives to antibiotic growth promoters (Partanen and Mrzoz, 1999). Their growth-promoting effects will generally be slightly less than that observed with antibiotics and they will be a more expensive solution. However given the current climate where antibiotic growth promoters are undesired at best and prohibited by law in many cases, organic acids have a useful role to play.

Improvement of phosphorus utilization in poultry has been demonstrated by addition of phytase to feeds (Chapter 4). Phytase has some limitations in terms of cost and heat stability and therefore alternative methods of increasing phosphorus utilization would be useful. One alternative may include organic acids, as citric acid in particular increases phosphorus utilization in maize-soybean diets and reduces the available phosphorus requirement by approximately 0.10% of the diet (Bolling-Frankenbach *et al.*, 2001).

There are several different acids and acid salts available and this makes their practical application somewhat confusing (Table 3). Generally the acids are all small molecules with molecular weights less than 200. Lactic, propionic, acetic, formic and phosphoric acids are all liquids in the pure state. The others are all solids. Phosphoric acid is an inorganic acid but is used in feeds together with the other organic acids.

Table 3 Various acids and acid salts used in feeds

Acetic	Fumaric
Benzoic	Lactic
Potassium benzoate	Calcium lactate
Sodium benzoate	Phosphoric
Sodium butyrate	Propionic
Citric	Ammonium propionate
Sodium citrate	Calcium propionate
Formic	Sorbic
Calcium formate	Potassium sorbate
Potassium formate	Tartaric

Acetic, sorbic and tartaric are not widely used in feeds; acetic because of its strong odour and flavour and sorbic and tartaric because of their high price. Benzoic acid and its salts are not permitted for feed use in the EU but can be used in other areas.

A further attraction of organic acids is that they are either of natural origin such as citric or lactic acids or are nature-identical in the case of formic, fumaric and propionic acids. Various organic acids are also widely used in the food industry as preservation agents or as acidulants for flavour and pH regulation. They are natural molecules that have little toxicity for higher animals including humans and they cannot leave any unnatural or toxic residues in food products. As indicated in Chapter 2 organic acids also play an important role in feed hygiene and in control of micro-organisms both in raw materials and in feeds. Organic acids are certainly a very important group of nutricines in Total Nutrition.

Probiotics

Manipulation of dietary components such as protein and various carbohydrates and the use of organic acid nutricines attempts to modulate the microflora in the gastrointestinal tract in a manner favourable to the health of the host animal. Another possible strategy is to exclude the growth of pathogens in the gastrointestinal tract by supplying to the animal a source of bacteria which will compete with the pathogens. This is the concept of "competitive exclusion." The general practice is to feed viable cultures of bacteria such as *Lactobacillus* species or *Bifidobacteriacea* species because these species of bacteria are widely considered to promote health and to inhibit growth of various pathogenic strains of bacteria existing in the gastrointestinal tract. Other bacterial species are also used as probiotics including *Saccharomyces* (yeast) species and the spore-forming organisms *Bacillus subtilis*.

Probiotic products have several possible modes of action. They may be able to directly inhibit or kill off the pathogenic bacteria and *Lactobacilli* species are often considered to be valuable here. A second mode of action is that of inhibiting the attachment of the pathogens to the wall of the gastrointestinal tract. Pathogenic bacteria in the gastrointestinal tract need to attach themselves to the wall of the gut to be able to develop disease syndromes and if probiotic cells can effectively compete with the pathogenic for binding sites or bind to the pathogens the virulence of the pathogens would be reduced.

There are however serious practical problems in their use in animal feeds. The major difficulty is that of heat stability and survival of the added bacteria in animal feeds after manufacture. Many bacterial species can be obtained as freeze-dried powders which when well packaged will have a long shelf-life so stability of probiotics products in itself should not be a cause for concern. The problem arises when these bacteria are added to feed mixtures, particularly those which will subsequently be pelleted. The pelleting process is recognised as having a beneficial effect in reducing the microbial load in feeds and consequently it is also likely to reduce the probiotic load.

Another practical difficulty with probiotics is that ideally they should establish themselves within the gastrointestinal tract but to do this they need to be of a suitable strain particular to the host animal and this is difficult in practice to achieve as standardised products must be commercially produced. A major difficulty with probiotics is that they only achieve a transient colonisation of the gastrointestinal tract. This could arise from a substantial loss of viability of the organisms on passage through the relatively hostile environment of the stomach and small intestine. The probiotic organisms would have to survive low pH and proteolytic enzymes.

The concept of competitive exclusion is quite attractive and consequently has been widely investigated and many proprietary probiotic products are available. One area perhaps where competitive exclusion has more chance of success, is in establishing a suitable microflora in the gastrointestinal tract of chicks to avoid *Salmonella* problems. Newly hatched chicks are particularly susceptible to intestinal colonization by *Salmonella* species. Resistance to this colonization by *Salmonella* rapidly increases with age as the normal gastrointestinal microflora become established (Nurmi and Rantala, 1973).

Complex cultures of micro-organisms have been developed based on the natural microflora of mature chickens and these are sprayed over day old chicks to inoculate them with the probiotic mixture. This procedure is not subjected to any heat processing as a feed application and therefore the bacterial population has a greater chance of surviving to establish itself in the gastrointestinal tract of the chick. Competitive exclusion using undefined cultures of normal gastrointestinal bacteria has given increased resistance to colonization by *Salmonella* in chickens raised under commercial conditions (Goren *et al.,* 1988).

Undefined cultures however raise questions about safety and methods of quality control although the undefined products are obtained from disease-free birds and absence of pathogens is always confirmed. Nevertheless the lack of knowledge about the role of the individual microbial components in a competitive exclusion mixture is certainly a problem in the current political climate.

The obvious solution to the problems of undefined cultures is the development of defined mixtures of micro-organisms. However this is not so simple and in many cases defined cultures are generally less effective than undefined cultures, although some success has been obtained with a culture of 29 strains of known caecal bacteria (Corrier *et al.,* 1995). Here treatment with the defined culture reduced the number of Salmonella-positive caecal contents in the processing plant from 5.7% in the control flocks to 2.7% in the treated flocks. However from a public health point of view an even greater reduction is required, ideally to zero.

A logical extension of the application of probiotics is to combine the bacterial preparation with carbohydrate nutricines and these mixtures have been termed "synbiotics." A combined treatment with a defined culture of caecal bacteria and lactose significantly decreased the number of *Salmonella* in the caecal contents of chicks. Furthermore there was an increased resistance by the chicks to colonization of the caecal and other organs by *S. enteritidis* (Corrier, *et al.,* 1994). However as lactose was incorporated into the feed at a level of 5% it would be difficult to use this in practice.

Another intriguing prospect for the application of probiotics is to bind mycotoxins and other toxic materials in the gastrointestinal tract of the animal. Mycotoxins, produced as a result of mould contamination and growth in feed raw materials, are a serious threat to animal health and productivity (Chapter 2). Several toxin binding products usually based on various minerals are available. Various *Lactobacilli* strains however also possessed mycotoxin-binding ability (Peltonen, *et al.*, 2000). It appears that specific probiotic lactic acid bacteria are able to bind mycotoxins to the bacterial cell wall and remove them from the gastrointestinal tract. This application of probiotics is not strictly a "probiotic effect" but more a chemical response as it appears that inactive bacterial cells can also bind the mycotoxins. This would overcome some of the problems discussed above in relation to heat stability of the live bacterial cells, although presumably feeding inactive bacterial cells would require a much higher dose rate than using live cultures which could multiply in the gastrointestinal tract. Nevertheless it is an extremely interesting possible application of probiotic micro-organisms.

Enteric disorders

Optimal health of the gastrointestinal tract of a food animal is crucial for maximum performance and for an economically viable animal production industry. The fundamental principle of animal production is converting feeds into human food products. As described in Chapter 4 the gastrointestinal tract plays a major role in this process and therefore any enteric disorders will have a dramatic effect upon animal productivity. The gastrointestinal tract of an animal is prone to infection by a whole host of pathogenic micro-organisms which will contribute to poor feed conversions and in some cases to increased mortalities. There are many connections between an animal's diet and the presence and extent of proliferation of pathogenic enteric bacteria. Feed components which are incompletely digested and absorbed in the small intestine and then enter the large intestine may predispose animals to certain diseases such as necrotic enteritis in poultry, swine dysentery in pigs, and laminitis in equines. Other problems are *Campylobacter* contamination and wet litter problems in poultry.

Control of these enteric diseases remains a major challenge in animal production. They are not always readily controlled by antibiotics or other medications and so application of various nutricines and careful diet formulation is also very important. The principles involved in control of enteric diseases may also apply to other bacterial populations in the gastrointestinal tract. For example, it may be possible to minimise the subclinical carriage of pathogenic bacterial species which can contaminate meat, and which present a threat to human health. *Campylobacter* contamination of broilers is an important issue but is difficult to control as it does not seem to be feed-borne although it is an organism which occurs in the gastrointestinal tract and therefore ought to be subject to control by dietary means.

Salmonella control in live animals

Young animals, and in particular newly hatched chicks, are very susceptible to *Salmonella* contamination and in serious cases high rates of infection occur with ensuing high levels of mortality. Fortunately resistance to *Salmonella* mediated disease increases with age as the normal intestinal flora becomes established.

Salmonella contamination of food animals has frequently been associated with poultry meat and eggs. However other species such as pigs can also carry *Salmonella*. In Denmark the recognition that *Salmonella* contamination of pork may pose a public health risk prompted the Danish Ministry of Agriculture to launch a nation-wide *Salmonella* control programme in the pig industry (Mousing *et al.*, 1997). Prevalence of *Salmonella* in pig herds has also been widely investigated in the Netherlands (van der Wolf *et al.*, 2001).

In this approach meat juice or blood samples from carcasses are examined for specific antibodies against *Salmonella* using an ELISA system. In the Danish system all herds are tested for *S.enterica* and are assigned into one of three categories. In level 1, the *S.enterica* prevalence is low and acceptable. Level 2 has moderate prevalence of *S. enterica*. Level 3 is unsatisfactory, usually >50% *S. enterica* prevalence for most herds in this category. If a herd is placed in Category 2 or 3 it must receive an advisory visit by a veterinarian and a pig specialist and improved hygiene practices must be undertaken. Pigs from herds in Category 3 must be slaughtered under special hygiene precautions.

There is significant interest today also in treating newly hatched chicks to protect them against infection by *Salmonella* using competitive exclusion based on probiotic products as described above. Newly hatched chicks are fed with anaerobic intestinal microflora obtained from healthy birds. The intestinal microflora from healthy adult birds helps to establish a beneficial microflora in the chick and to reduce the possibility of *Salmonella* contamination of the birds.

An ideal result from *Salmonella* control programmes in live animals would be to have 100% of the treated animals *Salmonella*-free. Unfortunately this is difficult to achieve but significant improvements can be made. Use of a known culture of caecal bacteria also gave some benefits but again not complete control (Corrier *et al.*, 1995). Nevertheless competitive exclusion techniques are currently in use and certainly offer an extra degree of protection against *Salmonella* contamination of live birds.

Another programme to reduce *Salmonella* in live birds is to feed sugars such as lactose or mannose as described above. This is an attractive prospect from the practical point of view as sugars are fairly stable molecules and could easily be incorporated into feeds. A disadvantage with this programme however is that large amounts of sugars need to be administered to birds either in feed or through the drinking water. In a recent study with *S. enteritidis* contamination of moulting layers, 2.5% lactose was supplied in the drinking water (Corrier, et al., 1994).

There is active research into possible methods to control *Salmonella* infections in live animals and dietary calcium phosphate offers some promise in this regard (Bovee-Ouderhoven, 1999). Dietary calcium phosphate in rats was able to increase the *Lactobacilli* and decrease the *Salmonella* in the ileum and in faeces (Table 4). The calcium phosphate is thought to interact with bile acids to reduce their cytotoxicity towards the *Lactobacilli* which in turn seems to reduce the invasion of tissues by *Salmonella* and helps the animal to resist colonization by *Salmonella*.

Table 4
Influeunce of calcium phosphate in the diet of rats on the number of *Lactobacilli* and *Salmonella* in the ileal contents and in the faeces ($\log_{10}$ CFU)

Bacteria	Location	Treatment	
		Control	Calcium phosphate
Lactobaccilli	Ileum	3.44a*	4.06b
Salmonella	Ileum	4.56a	3.34b
Lactobacilli	Faeces	7.68a	7.96b
Salmonella	Faeces	5.01a	4.20b

* Different letters in the same row are significantly different ($P<0.05$).

Vaccination of poultry against *Salmonella* is also becoming more prevalent. This has been used with laying hens in the USA (Holt, et al.,1996). Experimental evidence shows that *S. enteritidis* vaccination significantly decreased the progression of an *S. entertitidis* infection in hens, including the diminished production of eggs contaminated with *S. enteritidis*. Vaccination is claimed to play an important role in protecting the consumer against food-borne *Salmonella* infections. Vaccination will undoubtedly play an important role in *Salmonella* control in all animal species but is not yet fully developed.

Nectrotic enteritis

Necrotic enteritis in chickens, turkeys, and geese is caused by the bacteria *Clostridia perfringens* which is an important pathogen of poultry and also occurs in other species such as calves. This disease occurs sporadically in broilers and is very wide spread. It is a problem in Australia as well as in Europe. In broilers it frequently occurs between the ages of 2-6 weeks and can in severe cases lead to mortalities of 25-50 %. The disease causes a loss of appetite and the production of dark-coloured faeces. Post-mortem examination will show lesions in the wall of the small intestine. The wall may be congested and the gut lumen will frequently contain haemorrhagic material.

In healthy chickens clostridial bacteria normally live harmlessly in the lower gut and are found in the caeca and lower large intestine. The pH and high oxygen content of the healthy small intestine do not support growth of the organism. For disease symptoms to appear there needs to be a trigger factor that tips the balance in favour of the clostridial bacteria, allowing them to proliferate and migrate to the upper intestine. Several circumstances can predispose birds to necrotic enteritis; damage to the intestinal lining by coccidia or by other bacteria, immunosuppression and feed characteristics.

There has been some evidence that wheat and barley as major raw materials in poultry diets can increase the prevalence of necrotic enteritis in broilers compared to the use of maize-based diets (Branton, et al., 1987). The use of wheat in the diets influenced mortality in broilers (Table 5). Use of wheat as the only cereal in the diet increased mortality attributed to necrotic enteritis, by 6 to 10 times that experienced when corn was the only cereal used in the diet. Birds that consumed the diet containing approximately equal quantities of wheat and maize exhibited intermediate mortality.

Table 5
Effect of cereal type on performance of broilers at 42 days of age infected with necrotic enteritis

Cereal	Weight (kg)	FCR	Mortality (%)
Maize	1.749a	1.946a	12/420a (2.9)
Wheat	1.659a	1.861b	101/350 (28.9)
Maize/Wheat	1.757a	1.871b	44/350 (12.6)

The precise reason why wheat predisposes poultry to necrotic enteritis is not obvious. The water-soluble pentosans of wheat did not directly increase the growth of *C. perfringens* (Branton et al., 1996).) Supplementation of a diet containing wheat pentosans with the antibiotic procaine penicillin did not improve bird performance (Choct and Annison, 1992), and therefore why wheat stimulates problems of necrotic enteritis is still not clear.

Clostridial species are gram-positive bacteria and are not readily controlled by common non-medicinal feed additives such as organic acid mixtures. They are also fairly resistant to phenols, cresols and quaternary ammonia compounds used in various disinfectants. Most clostridia are susceptible to tetracyclines, penicillins and avilamycin. They seem to be more resistant to tilmicosin, tylosin and virginiamycin. Most strains isolated from chickens in the USA were resistant ot bacitracin and lincomycin (Watkins *et al.*, 1997). Amoxycillin in the drinking water is also an effective treatment for severe outbreaks of necrotic enteritis. A useful anticlostridial agent in the past has been avoparcin (Kaldhudal and Hofshagen, 1992). In a broiler trial, no birds showed symptoms of necrotic enteritis when offered feed supplemented with avoparcin. The recent banning of avoparcin from animal feeds has certainly not helped the clostridial problem.

There are a wide range of nutritional strategies in terms of feed modification which can positively impact upon the incidence of necrotic enteritis (Table 6).

Table 6
Reduction of the risk of necrotic enteritis by Total Nutrition

Feed modification	Expected benefit
Reduction in protein content through the use of synthetic amino acids	Reduces the amount of nitrogen which can enter the caeca
Use of whole grains or coarse particle sizes	Encourages the development of the gizzard and improves overall nutrient digestibility
Enzymes	Improve the digestion of nutrients in the small intestine and reduces the flow of nutrients into the large intestine.
Organic acids	Reduce the microbial load of the feed and limit fermentation in the small intestine
Prebiotics	Encourage the development of beneficial bacteria and limit the growth of pathogens.
Essential oils	Have an antibacterial effect and stimulate the secretion of digestive enzymes

A possible model for necrotic enteritis is that the initial stress might be the development of coccidiosis in the small intestine. This alters the absorptive surface and reduces the secretion of pancreatic enzymes. This leads to an excess of nutrients in the small intestine and these enter into the caeca inducing the rapid development of a number of micro-organisms including *C. perfingens* which rapidly multiplies and is then introduced into the small intestine where it alters the intestine by the secretion of toxins. Consequently treatments which improve the digestibility and absorption of nutrients should diminish the quantity of nutrients arriving in the large intestine and therefore reduce the risk of initiating necrotic enteritis.

Campylobacter in poultry

The genus *Campylobacter* is a recently recognised pathogen and was only proposed as a species in 1963. It is a rod-shaped Gram negative organism and is a thermophilic species capable of growing at a temperature as high as 42°C. It is a common micro-organism in the gastrointestinal tract of chickens but is not a pathogen for avian species. However in humans *Campylobacter* infection causes symptoms of headache, fever, vomiting and diarrhoea. These symptoms are similar to those caused by *Salmonella* but tend to be less severe. Most poultry-borne diseases are self-limiting and rarely last longer than one week.

The origin of infections by *Camplyobacter* is far from clear. Poultry flocks can certainly become infected from a contaminated environment. The existence of vertical transmission is still debatable and is not proven. Flocks are rarely contaminated early on in the rearing period and the reason for this is unknown. With broilers, development of *Camplyobacter* colonization in the gastrointestinal tract usually occurs only at two or three weeks of age.

A recent survey conducted in Germany (Atanossova and Ring, 1999) indicated very high levels of *Camplyobacter* contamination in poultry flocks, broiler carcasses and in wild pheasants (Table 7).

Table 7
Camplyobacter in poultry in Germany

Sample source	No. of samples	No. of Campylobacter positive	Campylobacter positive (%)
Poultry flocks	509	209	41.4
Broilers	111	51	45.9
Wild pheasants	52	14	25.9

It is not clear whether *Camplylobacter* comes from other animals or insects to broilers. However when intestinal colonization is established in a broiler flock enormous concentrations of the organism can develop in the gastrointestinal tract of the broiler. The *Camplylobacter* organism appears to establish a commensal relationship within the intestinal tract of birds. Excretion levels in excess of 10^8 cfu *Campylobacter*/g of bird faeces have been reported (Stern *et al.*, 1998). With such a large concentration of the organism clearly an enormous potential exists for the cross contamination of other flocks, domestic animals, rodents and insects. Sources of contamination and routes of infection in broiler flocks need to be controlled. Horizontal transmission from one living animal to another is of major concern.

Protective effects against *C. jejeuni* colonization in chickens have been obtained by feeding the sugar mannose (Schoeni and Wong, 1994) although a later study using mannanoligosaccharides from yeast did not show any response with several *Campylobacter* strains ((Spring et al., 2000). Treatment of drinking water for broilers with acetic, formic or lactic acids may be one possibility to reduce carcass contamination by *Campylobacter* (Byrd et al., 2001).

Campylobacter is of serious Public Health concern as it influences food safety. It will be difficult to control in broilers as it is not pathogenic and infected poultry show no clinical signs of disease. It will have to be controlled by dietary means but as it does not seem to be feed-borne simply treating feed with antibacterial organic acid mixtures is unlikely to be successful. Probably products will need to be added to feed or to drinking water which can inhibit *Campylobacter* in the gastrointestinal tract, and much work needs to be done to understand more about *Campylobacter* and diet. There are no obvious dietary stress factors yet recognised in connection with the development of *Campylobacter* in the gastrointestinal tract of chickens.

Wet litter in poultry

Poultry feed raw materials such as wheat and barley which have a high content of NSPs generate increased viscosity in the gastrointestinal tract of broilers and this has been associated with nutrient utilzation and performance problems. As indicated in Chapter 4 many of these problems have been solved by the addition of enzymes to feeds. One of the difficulties caused by feeding cereals high in NSPs to broilers is wet litter. This in effect seems to be an interaction between the microflora of the gastrointestinal tract and the feed components.

This is further confirmed by viscosity measurements. It is relatively easy to show that suspensions of wheat or barley form high-viscosity gels. However frequently the viscosity of the contents of the gastrointestinal tract when broilers are fed a wheat-based diet is much greater than that due to the wheat alone. Also gnobiotic or "germ-free" broilers do not generate high intestinal viscosity even when fed on wheat-based diets (Langhout, 1998). This clearly suggests there must be some interactions between feed components and the microflora to give this high viscosity.

Scouring in piglets and dysentery in pigs

Post weaning diarrhoea or collibacillosis is a perpetual problem in pig production and is an economically important enteric disorder worldwide. The condition is caused by the proliferation of haemolytic strains of *E. coli.* in the gastrointestinal tract and results in diarrhoea, dehydration, weight loss and even death in the initial weeks following weaning. Piglets are usually fed a good quality, highly digestible diet immediately after weaning in an attempt to avoid a post-weaning growth check. However if this feed is not adequately digested and absorbed it may support the proliferation of pathogenic bacteria in the gastrointestinal tract leading to scouring problems.

Proliferation of the pathogenic *E. coli* strains can be reduced by decreasing dietary protein levels (Goransson *et al.*, 1995). Also feeding high levels of soluble NSPs in the form of Guar gum resulted in higher levels of *E. coli* in the small intestine of piglets (McDonald *et al.*, 1999). There was also a clear reduction in weight gain when soluble NSP as Guar gum was incorporated into the feed for piglets. However inclusion of a another source of soluble NSPs, sugar-beet pulp to a cereal-based weaner diet did not adversely affect weight gain (Longland *et al.*, 1994) so the type of soluble NSP may also play a role. Soluble NSPs may well influence proliferation of pathogenic bacteria in several ways. Soluble NSPs certainly increase digesta viscosity and reduce nutrient diffusion to the wall of the gastrointestinal tract for absorption. This increased viscosity may well trap the bacteria and the substrates necessary for their growth in the lumen of the gastrointestinal tract.

An overall conclusion would suggests that formulating diets with reduced levels of protein and fermentable carbohydrates such as soluble NSPs or resistant starch would be beneficial in reducing the colonisation of haemolytic *E. coli* in the small intestine in piglets.

Incorporation of organic acids into piglet feeds is now a wide-spread practice, with the objective of reducing various scouring problems in the post-weaning phase. As discussed above various organic acids have an antibacterial effect and may reduce the pathogenic bacterial load in the gastrointestinal tract.

A common practice in many countries to control scouring in piglets is to use zinc oxide (ZnO) at 2.5-4.0 kg/tonne of feed and this is certainly an effective programme. Zinc oxide is highly insoluble and so little zinc is absorbed by the animals, most of it passes out in the faeces. This raises another important issue, that of soil and water pollution by the zinc oxide from the manure produced by the piglets.

Swine dysentery is an economically important disease of pigs in many parts of the world. The problem is caused by infection with the bacterium *Serpulina hyodysentariae* but the presence of the organism in pigs does not always lead to an expression of the disease. This is very similar to the situation observed in poultry with *Clostridium perfringens* and the expression of the disease, necrotic enteritis. For the development of both of these enteric diseases it seems that a stress of some sort is needed and this is probably related to the diet. As discussed above necrotic enteritis in poultry is strongly associated with wheat and barley-based diets.

Pigs fed a highly digestible diet based on cooked white rice and animal protein were completely protected against infection with a virulent strain of *S. hyodysenteriae* (Pluske *et al.*, 1996). Pigs fed diets based on steam-flaked maize and steam-flaked sorghum had a low incidence of swine dysentery where pigs fed on wheat or barley had high levels of disease expression. Pigs fed on oat chaff also did not develop clinical signs of swine dysentery (Pluske *et al.*, 1998). Oat chaff is a good source of insoluble NSPs (16%) and contains relatively little soluble NSPs (0.52%). This further confirms a possible link between readily fermentable carbohydrate and the onset of swine dysentery. In general it seems that problems of swine dysentery are reduced with diets containing low levels of soluble NSP and of resistant starch (Durmic *et al.*, 1998). However the traditional parameters of fermentation in the large intestine such as pH values of the digesta and volatile fatty acid concentration did not show any clear links with the observed reductions in swine dysentery. Nevertheless it is clear that a strategy based on reducing the quantity of fermentable carbohydrate entering the large intestine reduces the incidence of swine dysentery. This will need to be seriously considered in formulating feeds for pigs in Total Nutrition where diet must also maintain health and avoid disease.

Laminitis in horses

Laminitis, also known as founder in horses, may develop when the animals consume large amounts of rapidly fermentable carbohydrates in cereal grains or in pasture. The rapid fermentation of this carbohydrate in the large intestine of the horse results in a significant increase in lactic acid content in the digesta and a consequent decrease in pH. Lactic acid is produced by the proliferation of Gram-positive bacteria; *Streptococcus bovis* and *Lactobacillus* species.

Increased concentrations of lactic acid in the digesta also lead to a lowering in the pH of the faeces and this has been related to behavioural problems in horses such as wood chewing and the consumption of bedding (Willard *et al.*, 1977). This suggests a link between fermentation activity in the large intestine and health and behaviour in horses. Controlling the microflora in the gastrointestinal tract is still a major challenge for nutrition of all animals.

Future directions

Considerable research energy has been devoted to devising strategies for the manipulation of rumen fermentation, and this will undoubtedly continue. In monogastrics the gastrointestinal tract is a major consumer of dietary energy. More effective modulation of the growth and development of the gastrointestinal tract will be an important route to improve animal performance in the future. Modulation of the gastrointestinal tract is not only valuable in terms of digestion and absorption of nutrients but is also extremely important in controlling various enteric diseases. Antibiotic growth promoters have played a valuable role in both improving feed utilization and modulating the flora of the gastrointestinal tract. Total Nutrition will have to devise strategies based around the use of nutricines and novel feed ingredients which allow high levels of animal production in an economic manner.

The gastrointestinal tract is the largest organ in the body and consumes about 20% of dietary energy. Consequently efficient management of the gastrointestinal tract will have an important impact upon feed requirements of animals and this is turn has serious economic effects. There will be an increasing need to develop efficient systems of nutrient absorption so that nutrients can readily pass from the gut lumen into the tissues of the gastrointestinal tract and then on into the body of the animal to support growth. To some extent deposition of a large amount of tissue in the gastrointestinal tract structure and consumption of significant amounts of dietary energy can be considered as competing with production of food tissues. However it is far from obvious how the size and metabolic activity of the gastrointestinal tract can be regulated, although this could be a valuable economic objective.

The gastrointestinal tract is also an extremely complex interface between the animal and its environment. In physiological terms the gut lumen is outside the body of the animal and is a part of the external environment. It has to cope with a series of abrupt dietary changes at birth, at weaning and later during the growing phase. The modern requirement for reduction in use of drugs and medicines in animal production make it ever more important to understand the interactions between nutrition, gastrointestinal tract physiology, immunology and their influences upon health and productivity of animals. This will be of fundamental importance in designing new strategies to enhance disease resistance and to promote health through nutrition. A stable microflora is important in helping the animals resist enteric diseases and an imbalance in the microbial population of the gastrointestinal tract may predispose animals to health problems.

It is particularly important to maintain a stable microflora in the large intestine of monogastric animals. This is vital for conversion of undigested carbohydrates and provides some extra energy that the animals can use. In growing pigs and in sows in particular this can be quite significant and can amount to some 16% of the total energy supply to the animal (Mikkelsen and Jensen, 2000). With increased emphasis upon use of vegetable-based raw materials for feeds due to fears over BSE there will be more fibre used in animal feeds and it will be necessary to obtain perhaps even more energy for growth from fermentation in the large intestine.

Pigs in particular have a great capacity to utilise fibre as an energy source. Indeed they have a much greater ability to use dietary fibre than do humans and may obtain up to 30% of their maintenance energy requirement from volatile fatty acids produced in the large intestine (Varel and Yen, 1997). The large intestine of pigs contains all the predominant cellulose degrading bacteria normally found in the rumen of cattle. Unlike the rumen however the pig large intestine does not contain protozoa nor anaerobic fungi and considerably less methane gas is produced in pigs than in ruminants. Methane production in ruminants is generally considered a wasteful process and so pigs may be more efficient at fibre utilisation than ruminants in this respect.

High quality cereal grains as a feed raw material may become limiting in the future as animal production increases on a global scale. Also as previously mentioned the decline in use of animal fats due to concerns of BSE will increase demand for other energy sources. It will probably be necessary to develop feeding strategies for pigs based on maximising utilization of high fibre feeds. Given the inherent capacity of pigs to ferment dietary fibres there is probably much potential here to reduce feed costs and maintain suitable animal performance.

The resident microbial population in the gastrointestinal tract of monogastric animals is extremely diverse and complex. However the exact composition of the microflora in both pigs and poultry is still not fully known. This makes it difficult to understand how the different bacterial groups interact with dietary components and how and when they may proliferate and cause enteric disorders. This lack of information makes it difficult to predict how a specific nutrient of nutricine in the feed will influence any specific pathogen. Clearly much more basic information on the microbiological composition of the gastrointestinal tract is an important requirement to enable feeds to be designed to assist with the control of enteric disorders.

Other characteristics of the digestive process also undoubtedly influence the colonisation by pathogenic bacteria as discussed above. The amount of digested nutrients remaining in the gastrointestinal tract that act as a substrate for the growth of pathogens is important as is the intestinal viscosity. Nevertheless it is clear that feed components can influence the occurrences of enteric diseases. Modulation of the gastrointestinal microflora to enhance overall animal health may be achieved by feeding specific NSPs and the use of various supplementary enzymes and perhaps liquid feeding.

Feeding animals non-digestible but fermentable carbohydrates such as oligofructans or resistant starch seems to significantly lower serum triglyceride and phospholipid levels (Delzenne and Kok, 1998). These carbohydrates can exert a systemic effect and influence liver metabolism to reduce lipid synthesis.

These observations are of great importance for future developments in Total Nutrition. The presence of certain carbohydrate nutricines in the diet and the avoidance of oxidative stress on the animal may influence changes in the virulence of various pathogens and upon lipid metabolism. If this can be further developed it will be an important advantage to use nutritional strategies to control pathogens. This phenomenon has enormous implications for animal health. There is still much to learn about the various physiological roles of carbohydrates in nutrition.

An interesting possible application of carbohydrate nutricines may be to attenuate or reduce the virulence of certain pathogens. Fore example the pathogenicity of *Listeria monocytogenes* is repressed in the presence of the carbohydrate derived from celluose known as cellobiose (Park and Kroll, 1993). This micro-organism is not virulent in its natural habitat in the soil where it is in contact with decomposing plant materials which would include cellobiose. In humans however the absence of free cellobiose allows virulence to be expressed and the micro-organism becomes pathogenic. If similar situations were found for other pathogens it would suggest that various carbohydrate nutricines could exert an important effect in controlling the activity of pathogens.

It is known that the virulence of some viruses is related to the nutritional status of an animal (Beck, 1999). A normally-benign strain of coxsackievirus B3 becomes virulent in selenium or vitamin E deficient mice (Beck, 1999). This change in virulence is due to specific mutations in the virus itself such that once the mutation has occurred even mice with normal nutrition become vulnerable to the virus. Studies with coxsackievirus in mice indicated that oxidative stress in the host animals leads to mutations in the virus which in turn leads to increased virulence of the particular virus. A virulent stain of coxsckievirus became even more virulent under oxidative stress.

A similar phenomenon has been observed in pigs with porcine circovirus 1 which is non-pathogenic but which has mutated into porcine circovirus 2 which is highly pathogenic.

Campylobacter contamination of broilers is a particularly intractable problem as the organism does not seem to be feed-borne. Consequently good feed hygiene, important as it is, will not likely solve the problem of *Campylobacter*. It will be necessary to develop some feed component which could modulate *Campylobacter* in the gastrointestinal tract. Initial results with non-digestible carbohydrates are not encouraging (Spring *et al.*, 2000), although further work would be warranted in this area. Some organic acids such as lactic and fumaric are slowly absorbed from the gastrointestinal tract and these may also have a role to play in control of *Campylobacter*. There is also the potential to develop competitive exclusion products targeted at *Campylobacter* (Schoeni and Wong, 1994).

There is probably considerable scope for further development in the application of organic acids in animal nutrition. In addition to their antimicrobial effects they also may improve utilzation of various nutrients such as phosphorus (Bolling-Frankenbach, 2001).

An assessment of the fermentability of various feed ingredients will be needed to formulate feeds either high or low in fermentable substrates. The fermentation characteristics of soyabean fractions have already been studied using an *in vitro* cumulative gas production technique (van Laar *et al.*, 2000). In this procedure fermentable substrates are incubated with an inoculum prepared from pig faeces and gas production monitored as an indication of fermentability of the various substrates. This procedure could be applied to many feed ingredients to determine their content of fermentable material.

Much of the information available on the microflora of the gastrointestinal tract has been obtained through classic microbiology and culture of a wide range of different organisms. However there are serious limitations to the application of the classic techniques to such complex systems as the microflora of the gastrointestinal tract. Many micro-organisms present in the gastrointestinal tract are strict anaerobes for example and are not easy to isolate and to culture. The huge bacterial population in the gastrointestinal tract makes characterisation by classic microbiology extremely labour intensive. However now there is increasing use of molecular biology techniques to characterise the microflora, to monitor population dynamics and to track specific strains of bacteria in the gastrointestinal tract (Steer *et al.*, 2000). These new techniques will greatly increase our understanding of the gastrointestinal tract and its role in health, nutrition, and disease.

There will be a greater interest in the use of diet to control enteric disorders as we move further away from routine use of antibiotics. There will be an increasing interest in the development of non-therapeutic disease management which will probably be strongly focused on disease avoidance as a strategy. Another important aspect of this would be to control intestinal pathogens which can contaminate meat and which present a threat to human health. *Salmonella* and *Campylobacter* species in meat products are already significant public health issues. The possibility to formulate feeds with appropriate nutrients and nutricines to control animal health problems and also the quality of the final human food products is a very attractive goal for future development.

The weaning process in pigs still requires optimisation and perhaps this can be accomplished by developing feed formulations which promote a favourable bacterial profile whilst also enhancing the production of SCFAs, especially butyric acid. Various NDOs and resistant starches could assist by enhancing the generation of a favourable SCFA profile in the large intestine.

Vaccination techniques undoubtedly have a solid future in protecting animals from enteric diseases. Vaccines against *Salmonella* are already available and in the future vaccines against many other enteric pathogens will likely be developed. Combination of effective vaccines and diet manipulation, taking advantage of various organic acid and carbohydrate nutricines should enable animals to avoid many enteric diseases.

References

Adams, C. A. (1999). *Nutricines. Food Components in Health and Nutrition*. Nottingham University Press, UK. pp. 57-73.

Atanassova, V. and Ring, C. (1999). Prevalance of *Campylobacter* spp. in poultry and poultry meat in Germany. *International Journal of Food Microbiology*, **51**, 187-190.

Bardócz, S., Grant, G. and Pusztai, A. (1998). Why do we need polyamine transgenic plants. In: *Biogenically Active Amines in Food*. Eds: S. Bardòcz, A. white and A. F. Tiburcio. COST 917, European Commission, pp.1-8.

Barnes, E. M., Impey, C. S. and Stevens, B. J. H. (1979). Factors affecting the incidence and anti-salmonella activity of the anaerobic caecal flora of the young chick. *Journal of Hygiene*, **82**, 263-283.

Beck, M. A. (1999). Selenium and host defence towards viruses. *Proceedings of the Nutrition Society*, **58**, 707-711.

Blank, R., Mosenthin, R., Sauer, W. C. and Huang, S. (1999). Effect of fumaric acid and dietary buffering capacity on ileal and fecal amino acid digestibilities in early-weaned pigs. *Journal of Animal Science*, **77**, 2974-2984.

Bolduan, G., Jung, H., Schnabel, E. and Schneider, R. (1988). Recent advances in the nutrition of weaner piglets. *Pig News and Information*, **9**, 381-385.

Bolling-Frankenbach, S. D., Snow, J. L., Parsons, C. M. and Baker, D. H. (2001). The effect of citric acid on the calcium and phosphorus requirements of chicks fed corn-soybean meal diets. *Poultry Science*, **80**, 783-788.

Bovee-Oudenhoven, I. M., Wissink, M. L., Wouters, J. T. and Van der Meer, R. (1999). Dietary calcium phosphate stimulates intestinal lactobacilli and decreases the severity of a salmonella infection in rats. *Journal of Nutrition*, **129**, 607-612.

Branton, S. L., Reece, F. N. and Hagler Jr, W. M. (1987). Influence of a wheat diet on mortality of broiler chickens associated with necrotic enteritis. *Poultry Science*, **66**, 1326-1330.

Branton, S. L., Lott, B. D., May, J. D., Hedin, P. A., Austin, F. W., Latour, M. A. and Day, E. J. (1996). The effects of nonautoclaved and autoclaved water-soluble wheat extracts on the growth of *Clostridium perfringens*. *Poultry Science*, **75**, 335-338.

Byrd, J. A., Hargis, B. M., Caldwell, D. J., Bailey, R. H., Herron, K. L., McReynolds, J. L., Brewer, R. L., Anderson, R. C., Bischoff, K. M., Callaway, T. R. and Kubena, L. F. (2001). Effect of lactic acid administration in the drinking water during preslaughter feed withdrawal on Salmonella and Campylobacter contamination of broilers. *Poultry Science*, **80**, 278-283.

Cant, J. P., McBride, B. W. and Croom, W. J. (1996). The regulation of intestinal metabolism and its impact on whole animal energetics. *Journal of Animal Science*, **74**, 2541-2553.

Choct, M. and Annison, G. (1992). Anti-nutritive effect of wheat pentosans in broiler chickens: role of viscosity and gut microflora. *British Poultry Science*, **33**, 821-834.

Corrier, D. E., Nisbet, D. J., Scanlam, C. M., Tellez, G., Hargis, B. M. and DeLoach, J. R. (1994). Inhibition of *Salmonella enteritidis* cecal and organ colonization in leghorn chicks by a defined culture of cecal bacteria and dietary lactose. *Journal of Food Protection*, **56**, 377-381.

Corrier, D. E., Nisbet, D. J., Scanlam, C. M., Hollister, A. G., Caldwell, D. J., Thomas, L. A., Hargis, B. M., Tomkins, T. and Deloach, J. R. (1995). Treatment of commercial broiler chickens with a characterised culture of cecal bacteria to reduce salmonellae colonization. *Poultry Science*, **74**, 1093-1101.

Delzenne, N. M. and Kok, N. (1998). Effect of non-digestible fermentable carbohydrates on hepatic fatty acid metabolism. *Biochemical Society Transactions*, **26**, 228-230.

Durmic, Z., Pethick, D. W., Pluske, J. R. and Hampson, D. J. (1998). Changes in bacterial populations in the colon of pigs fed different sources of dietary fibre, and the development of swine dysentry after experimental infection. *Journal of Applied Microbiology*, **85**, 574-582.

Ewing, W. N. and Cole, D. J. A. (1994). *The Living Gut*. Context, UK. pp. 9-28.

Gálfi, P. and Neogrády, S. (2001). The pH-dependent inhibitory action of n-butyrate on gastrointestinal epithelial cell division. *Food Research International*, **34**, 581-585.

Geary, T. M., Brooks, P. H., Beal, J. D. and Campbell, A. (1999). Effect on weaner pig performance and diet microbiology of feeding a liquid diet acidified to pH 4 with either lactic acid or through fermentation with Pediococcus acidilactici. *Journal of the Science of Food and Agriculture*, **79**, 633-640.

Gibson, G. R. and McCartney, L. A. (1998) Modification of the gut microflora by dietary means. *Biochemical Society Transactions*, **26**, 222-228.

Gibson, G. R. and Roberfoid, M. B. (1995). Dietary modulation of the human colonic microbiota. *Journal of Nutrition*, **125**, 1401-1412.

Goransson, L., Lange, S. and Lonnoroth, I. (1995). Post weaning diarrhoea: focus on diet. *Pig News and Information*, **16**, 89N-91N.

Goren, E., de Jong, W. A., Doornenbal, P., Bolder, N. M., Mulder, R. W. A. W. and Jensen, A. (1988). Reduction of *Salmonella* infection of broilers by spray application of intestinal microflora; a longitudinal study. *Veterinary Quarterly*, **10**, 249-255.

Hillman, K., Murdoch, T. A., Spencer, R. J. and Stewart, C. S. (1994). The inhibition of enterotoxigenic *Escherichia coli* by the microflora of the porcine ileum in an *in vitro* semicontinuous culture system. *Journal of Applied Bacteriology*, **76**, 294-300.

Holt, P. S. Stone, H. D., Gast, R. K. and Porter, Jr., R. E. (1996). Growth of Salmonella enteritidis (SE) in egg contents from hens vaccinated with an SE bacterin. *Food Microbiology*, **13**, 417-426.

Ijl, P. A. and Tivey, D. R. (1989). Natural and synthetic oligosaccharides in broiler chicken diets. *World's Poultry Science Journal*, **54**, 129-143.

Kaldhusdal, M. and Hofshagen M. (1992). Barley inclusion and avoparcin supplementation in broiler diets. 2. Clinical, pathological, and bacteriological findings in a mild form of necrotic enteritis. *Poultry Science*, **71**, 1145-1153.

Kleessen, B., Stoof, G., Proll, J., Schmiedl, D., Noack, J. and Blaut, M. (1997). Feeding resistant starch affects fecal and cecal microflora and short-chain fatty acids in rats. *Journal of Animal Science*, **75**, 2453-2462.

Langhout, D. J. (1998). *The Role of the Intestinal Flora as Affected by Non-starch Polysaccharides in Broiler Chicks*. Ph. D. Thesis, Agricultural University of Wageningen, Netherlands.

Longland, A. C., Carruthers, J. and Low, A. G. (1994). The ability of piglets 4 to 8 weeks of age to digest and perform on diets containing two contrasting sources of non-starch polysaccharides. *Animal Production*, **58**, 405-410.

Mathew, A. G., Franklmin, M. A., Upchurch, W. G. and Chattin, S. E. (1996). Effect of weaning on ileal short-chain fatty acid concentrations in pigs. *Nutrition Research*, **16**, 1689-1698.

McDonald, D. E., Pethick, D. W., Pluske, J. R. and Hampson, D. J. (1999). Adverse effects of soluble non-starch polysaccharide (guar gum) on piglet growth and experimental colibacillosis immediately after weaning. *Research in Veterinary Science*, **67**, 245-250.

Mikkelsen, L. L. and Jensen, B. B. (2000). Effect of fermented liquid feed on the activity and composition of the microbiota in the gut of pigs. *Pig News and Information*, **21**, 59N-66N.

Mousing, J., Thode Jensen, P., Halgaard, C., Bager, F., Feld, N., Nielsen, B., Nielsen, J. P., and Bech-Nielsen, S. (1997). Nation-wide *Salmonella enterica* surveillance and control in Danish slaughter swine herds. *Preventive Veterinary Medicine*, **29**, 247-261.

Nurmi, E. and Rantala, M. (1973). New aspects of *Salmonella* infection in broiler production. *Nature*, **241**, 210-211.

Overland, M., Granli, T., Kjos, N. P., Fjetland, O., Steien, S. H. and Stokstad, M. (2000). Effect of dietary formates on growth performance, carcass traits, sensory quality, intestinal microflora, and stomach alterations in growing-finishing pigs. *Journal of Animal Science*, **78**, 1875-1884.

Oyofo, B. A., Drolesky, R. E., Norman, J. O., Mollenhauer, H. H., Ziprin, R. L., Corrier, D. E. and DeLoach, J. R. (1989a). Inhibition by mannose of *in vitro* colonization of chicken small intestine by *Salmonella typhimurium*. *Poultry Science*, **68**, 1351-1356.

Oyofo, B. A., DeLoach, J. R., Corrier, D. E., Norman, J. O., Ziprin, R. L. and Mollenhauer, H. H. (1989b). Effect of carbohydrates on *Salmonella typhimurium* colonization in broiler chickens. *Avian Diseases*, **33**, 531-534.

Park, S. F. and Kroll, R. G. (1993). Expression of listeriolysin and phosphatidylinositol-specific phospholipase C is repressed by the plant-derived molecule cellobiose in *Listeria monocytogenes*. *Molecular Microbiology*, **8**, 653-661.

Parks, C. W., Grimes, J. L., Ferket, P. R. and Fairchild, A. S. (2001). The effect of mannanoligosaccharides, bambermycins, and virginiamycin on performance of large white male market turkeys. *Poultry Science*, **80**, 718-723.

Partanen, K. H. and Mrzoz, Z. (1999). Organic acids for performance enhancement in pig diets. *Nutrition Research Reviews*, **12**, 117-145.

Peltonen, K. D., El-Nezami, H. S., Salminem, S. J. and Ahokas, J. T. (2000). Binding of aflatoxin B_1 by probiotic bacteria. *Journal of the Science of Food and Agriculture*, **80**, 1942-1949.

Pluske, J. R., Siba, P. M., Pethick, D. W., Durmic, Z., Mullan, B. P. and Hampson, D.J. (1996). The incidence of swine-dysentery in pigs can be reduced by feeding diets that limit the amount of fermentable substrate entering the large intestine. *Journal of Nutrition*, **126**, 2920-2933.

Pluske, J. R., Pethick, D. W. and Mullan, B. P. (1998). Differential effects of feeding fermentable carbohydrates to growing pigs on performance, gut size and slaughter characteristics. *Animal Science*, **67**, 147-156.

Pluske, J. R., Durmic, Z., Pethick, D. W., Mullan, B. P. and Hampson, D. J. (1998). Confirmation of the role of readily fermentable carbohydrates in the expression of swine dysentery in pigs after experimental infection. *Journal of Nutrition*, **128**, 1737-1744.

Reid, C. A. and Hillman, K. (1999).The effects of retrogradation and amylose/amylopectin ratio of starches on carbohydrate fermentation and microbial populations in the porcine colon. *Animal Science*, **68**, 503-510.

Risley, C. R., Kornegay, E. T., Lindemann, M. D. and Weakland, S. M. (1991). Effects of organic acids with or without a microbial culture on performance and gastrointestinal tract measurements of weanling pigs. *Animal Feed Science and Technology*, **35**, 259-270.

Schoeni, J. C. L. and Wong, A. C. L. (1994). Inhibition of *Campylobacter jejuni* colonization in chicks by defined competitive exclusion cultures (S-layer). *Applied Environmental Microbiology*, **60**, 1191-1197.

Spring, P., Wenk, C., Dawson, K. A. and Newman, K. E. (2000). The effects of dietary mannanoligosaccharides on cecal parameters and the concentrations of enteric bacteria in the ceca of Salmonella-challenged broiler chicks. *Poultry Science*, **79**, 205-211.

Stern, N. J., Cox, N. A., Line, J. E. and Bailey, J. S. (1998). *Safe Chicken for the Next Century*. European Commission pp 69-74.

Steer, T., Carpenter, H., Tuohy, K. and Gibson, G. R. (2000). Perspectives on the role of the human gut microbiota and its modulation by pro- and prebiotics. *Nutrition Research Reviews*, **13**, 229-254.

Van Der Meulen, J and Jansman, A. J. M. (1997). Nitrogen metabolism in gastrointestinal tissue of the pig. *Proceedings of the Nutrition Society*, **56**, 535-545.

Van der Wielen, P. W. J. J., Biesterveld, S., Notermans, S., Hofstra, H., Urlings, B. A. P. and van Knapen, F. (2000). Role of volatile fatty acids in development of the cecal microflora in broiler chicks during growth. *Applied and Environmental Microbiology*, **66**, 2536-2540.

Van der Wolf, P. K., Wolbers, W. B., Elbers, A. R. W., van der Heijden, H. M. J. F., Koppen, J. M. C. C., Hunneman, W. A., van Schie, F. W. and Tielen, M.J. M. (2001). Herd level husbandry factors associated with the serological *Salmonella* prevalence in finishing pig herds in the Netherlands. *Veterinary Microbiology*, **78**, 205-219.

Van Laar, H., Tamminga, S., Williams, B. A., Verstegen, M. W. A. and Schols, H. A. (2000). Fermentation characteristics of polysaccharide fractions extracted from the cell walls of soya bean cotyledons. *Journal of the Science of Food and Agriculture*, **80**, 1477-1485.

Varel, V. H. and Yen, J. T. (1997). Microbial perspective on fibre utilization by swine. *Journal of Animal Science*, **75**, 2715-2722.

Watkins, K. L., Shyrock, T. R., Dearth, R. N. and Y. M. Saif. (1997). In-vitro antimicrobial susceptibility of *Clostridium perfringens* from commercial turkey and broiler chicken origin. *Veterinary Microbiology*, **54**, 195-200.

Willard, J. G., Willard, J. C., Wolfram, S. A. and Baker, J. P. (1977). Effect of diet on caecal pH and feeding behaviour of horses. *Journal of Animal Science*, **45**, 87-93.

6   External Enemies: Immune System and Defence in a Dangerous World

Animals inevitably live in a hostile environment and they must continuously struggle against the hazards of climate, toxic materials, predators, and various pathogenic micro-organisms. Animals are exposed to an enormous range of micro-organisms, some of which readily penetrate the epithelia of the respiratory, gastrointestinal and genital systems and invade the body. Many micro-organisms live in the gastrointestinal tract and other cavities of most animals. As discussed in Chapter 5 the huge bacterial population in the gastrointestinal tract can be beneficial or deleterious to the health of the animal and must be well-managed. Virus infections of the gastrointestinal tract occur commonly in pigs and poultry and can cause little or no symptoms, or they can lead to catastrophic losses. Furthermore virus infections of the gastrointestinal tract are likely to encourage the development of other diseases. Viruses may damage the mucosal lining of the gastrointestinal tract and provide an access point for other potential pathogens such as *E. Coli* or *Salmonella* spp. Such damage may also allow attachment of other pathogens to the wall of the gastrointestinal tract. Damage to the gastrointestinal tract and the onset of diarrhoea syndromes caused by various pathogens will also have a secondary effect upon the ability of the infected animal to digest and absorb nutrients. Growth of pathogenic micro-organismsm must be controlled to avoid direct disease syndromes, opportunistic infections and severe enteric disorders.

Figure 1
Effects of infection by pathogenic organisms which result in reduced growth rates of animals

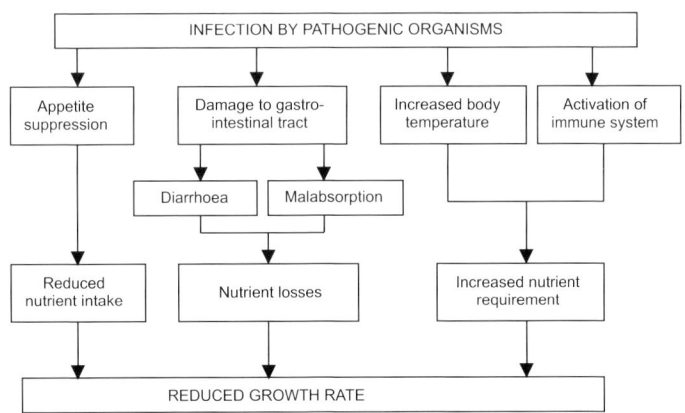

Infection by disease-causing organisms has several important consequences directly related to nutrition as illustrated in Figure 1. Suppression of appetite will reduce overall nutrient intake which is of major importance in maintenance of good animal growth. Damage to the tissues of the gastrointestinal tract may occur, either to the villi or possibly through development of lesions such as in necrotic enteritis in poultry. Such damage is usually manifested as diarrhoea and malabsorption of nutrients from the feed. The overall result is a loss of ingested nutrients which should be used for growth. Increased body temperature and activation of the immune system increases the nutrient requirement, but these nutrients are not used to support growth of the animal body. Infection by pathogenic organisms has a serious effect upon the productivity of animals and therefore in Total Nutrition much emphasis is placed upon disease avoidance to reduce infections to the minimum.

Despite the serious and persistent challenges of the environment to health, animals survive remarkably well due to a range of defence mechanisms against environmental hazards. These defence mechanisms involve non-specific physical barriers, a chemical defence system based upon enzymes, acids and biologically active peptides and a physiological response known as the immune system (Table 1).

Table 1
Defence systems against microbial diseases

Type of defence system	Components of defence system
Physical	Skin, feathers, hair, epithelial layers of gastrointestinal and respiratory tracts
Chemical	Acids: HCl, fatty acids, organic acids Enzymes: lysozyme, lactoperoxidase, pepsin Cytochrome P450 oxidase system, Peptides
Physiological (cellular and humoral immunity)	Innate immune system: macrophages, neutrophils Adaptive immune system: T and B lymphocytes

Maintenance of all these systems is influenced by nutrition. Stimulation of the immune system by various feed components is now of great interest and is described as nutritional immunology or immunonutrition (Grimble, 2001). A reduced immune function is often seen in animals under stress or after exposure to pathogenic organisms. Immunostimulation is essential to restore a suboptimal immune system.

Therefore Total Nutrition must also positively focus upon supporting and strengthening the various defence systems of the animal. This becomes even more important when reliance upon antibiotics and medications to defend the animal against environmental hazards is no longer readily available.

Physical barriers

Animals possess non-specific physical barriers such as feathers, skin, and hair which protect the body tissues against injury and prevent pathogens gaining access to the body. Damage to these protective barriers will allow easy access of pathogens into the body. Feather pecking in poultry and fighting and tail biting in pigs can cause serious damage to the skin. Wherever possible diets need to be formulated to maintain good feather and skin development for the physical protection of the body.

The mucosa, or epithelial surfaces, of the respiratory system and of the gastrointestinal tract is the first line of defence from harmful pathogens and antigens in the air or ingested in feed. The epithelia of the gastrointestinal tract must have a good physical integrity to prevent the rapid transport of pathogens into the body. Yet it must be sufficiently thin to actively transport nutrients. These epithelial surfaces can be damaged in various ways. High ammonia content in the housing systems damages lung tissue. Diet changes and weaning stress may damage the epithelial layers in the gastrointestinal tract and make the young animals more susceptible to pathogens. In weaned piglets reduction of feed intake leads to rapid atrophy of the villi in the gastrointestinal tract (Pluske *et al.*, 1996). Consequently it is important that diets and feeding system for young piglets are designed to ensure continuous feed intake.

Chemical defence system

Once pathogens have entered the body either in feed, from the environment or as a result of injury they are dealt with initially by a chemical defence system,. This has several components: acids, enzymes and peptides. Pathogens entering the stomach have to resist the very low pH due to production of HCl. It is generally accepted that some enteric problems with young piglets are due to a lack of secretion of HCl by the immature stomach and this makes the animal more susceptible to enteric diseases. This has led to the practice of including various organic acid mixtures in piglet feeds (Partanen and Mroz, 1999).

Several different fatty acids have good antimicrobial activity as described in Chapter 2 and play an important role in feed hygiene as well as influencing the situation in the gastrointestinal tract. The medium chain fatty acid, lauric acid has good activity against Gram-positive organisms (Conley and Kabara, 1973). Various organic acids will play an important role in diet formulation in the absence of antibiotics.

Various enzymes also play a role in defence systems. Lysozyme occurs in tears, sweat, gastric juice, saliva and egg white. It is particularly effective against Gram-positive bacteria. Lactoperoxidase occurs in milk and has an antibacterial effect. Pepsin is a protein-digesting enzyme that acts at a low pH. The action of pepsin and the low pH in the stomach keep the bacterial population low in the stomach.

Cytochrome P450 is a mixed function oxidase system. The main function of these enzymes is to metabolize toxic materials usually by hydroxylation or oxidative reactions. This increases the water solubility of the toxic molecules and facilitates their excretion from the body.

All vertebrates from amphibians to humans seem to have developed a defence system based on a series of broad-spectrum antimicrobial peptides also known as bio-active peptides (Nicolas and Mor, 1995). These peptides are secreted by cells of the mucosal surfaces of the respiratory and gastrointestinal tracts as well as by granular glands of the skin. They cause the lysis of numerous pathogenic micro-organisms, including viruses, Gram-negative and Gram-positive bacteria, protozoa, yeasts and fungi.

These peptides are relatively small (20-46 amino acids) basic (lysine or arginine-rich) and amphipathic. Their antimicrobial activity most likely results from their capacity to form channels or pores within the microbial membrane in order to permeate the cell and impair its ability to carry out essential metabolic processes. This chemical defence system provides the animal with a repertoire of small peptides that are promptly synthesised upon induction, easily stored in large amounts and readily available to combat invading pathogens. They can be produced much faster than the rate at which micro-organisms proliferate. This chemical defence system may also be important in case of failure of the classical immune system as in immunosuppressed animals.

Immune system

All vertebrates possess an immune system that is designed to recognise and selectively eliminate invading pathogens. The immune system is able to recognise the presence of macromolecules of non-self origin. The immune system is extremely complex and is probably the second most complex system of the body after the brain. It consists of different organs and cell types. The primary organs are the thymus and bone marrow in mammals and the thymus and bursa of Fabricius in avian species. Secondary organs of the immune system include the spleen, various lymph nodes and Peyer's patches. It is still not completely understood how the immune system recognises and repels harmful pathogens including ones that the body has not encountered before, while tolerating the billions of beneficial bacteria in the gastrointestinal tract. It is able to recognise and destroy dangerous toxins while allowing the digestion and absorption of vital nutrients.

An important feature of the immune system is also its ability to retain a "memory" of the presence of a specific pathogen so that subsequent infection by the same pathogen will elicit a more rapid response. This "memory" of the immune system is the basis of vaccination where an animal is deliberately exposed to a safe form of a pathogen. On a subsequent exposure to the virulent pathogen the "memory" function of the immune system elicits a protective response against the pathogen.

The immune system in both mammals and in avian species comprises two basic systems: the innate system and the adaptive system. Innate immunity is thought to have evolved earlier than the adaptive system. It occurs in all multicellular organisms whereas the adaptive system is found only in vertebrates. The innate system is located in cells of the lymphatic system whilst the adaptive immunity resides ultimately in the blood serum. The two immune systems are distinct but are also integrated to provide overlapping protection. The innate system initiates and regulates the adaptive response. There are two functional components of the immune system, a cell-mediated or phagocytic system and a humoral or antibody mediated system.

The inner surface of the gastrointestinal tract in animals has an enormous surface area and more than 70% of the total amount of immune cells of the body are found in the gut wall. The first major encounter of feed components with the immune system occurs in the gastrointestinal tract. Consequently nutrition and immune status are closely linked. Harmful materials, known as antigens, which include bacteria, viruses and pesticides are recognised as distinct from innocuous feed proteins. In the case of microbial antigens a full activation of the immune system occurs if the defences of the gastrointestinal tract are breached. However in the case of feed proteins it is important to suppress this immune response and allow feed proteins to be digested and absorbed from the gastrointestinal tract.

When micro-organisms succeed in penetrating the epithelial layers of the gastrointestinal or respiratory tracts they initially encounter the innate immune system which is the first line of immune defence to limit the spread of an infectious challenge. In mammals, cells known as macrpohages, neutrophils, and natural killer cells constitute a major part of the innate immune defences against invading micro-organisms. Macrophages are mobile scavenger cells and are widely distributed throughout the body. They are phagocytic in nature and bind, engulf and degrade foreign antigens such as bacteria without any lag period after encounter. In poultry heterophils are the equivalent of neutrophils in mammals. Heterophils are especially important in young chickens that have not yet developed an acquired immunity to pathogens.

Components of the innate immune system are able to recognize structures on a broad range of micro-organisms and rapidly kill the invading pathogens. Cells of the innate immune system seem able to recognise various molecular structures that are shared by groups of pathogens. For example molecules such as lipolysaccharides and teichoic acids are the recognition structures for Gram-negative and Gram-positive bacteria respectively, double stranded RNA is involved in virus recognition and mannan-oligosaccharides are recognised in yeasts. They respond very rapidly to an immunological challenge to manifest the first line of the immune defence against infections. Foreign antigens such as pathogenic bacteria are readily degraded by enzymes and reactive oxygen species produced by the macrophages. Natural killer cells have a spontaneous cytotoxicity against various target cells such as malignant cells or normal cells infected with viruses, fungi and parasites. The innate immune system is rapid in action and most invading pathogens will be destroyed within a few hours.

The innate immune system also has another component termed complement. This is a defence system consisting of an enzymatic cascade of over 30 serum proteins normally present in blood in an inactive state. The cascade can be triggered either by the presence of cell walls of micro-organisms or by the binding of antibodies to the surface of the micro-organism.

If invading micro-organisms survive the innate immune system then the adaptive response is activated. Adaptive immunity is mediated by cells known as lymphocytes and these cells are exclusively involved in the specific immune recognition of foreign substances. The cell surfaces of the lymphocytes carry membrane bound antibodies and antibody-like molecules that function as antigen receptors and they also produce soluble antibodies. The lymphocytes are capable of generating several billion different antigen receptors or antibodies and this endows the animal with the capacity to recognise any antigen that it may encounter in its lifetime.

Lymphocytes become active after encountering specific antigens and comprise T cells and B cells. T cells are further classified into CD4$^+$ and CD8$^+$ lymphocytes. After activation these differentiate into a host of various effector T cells with different functions. Some of these T cells are cytotoxic and their function is to kill host cells that are infected by intracellular micro-organisms such as viruses. Other T cells (T-helper) known as TH1 and TH2 cells influence macrophages and so interact with the innate immune system. Individual B cells, when activated by recognition of a foreign invader differentiate into plasma cells which produce antibodies, serum proteins, that bind specifically to foreign substances and initiate a variety of elimination responses. These plasma cells produce humoral antibodies with the same antigen-binding sites as their cell-surface receptors. In the gastrointestinal tract B cells produce mainly immunoglobulin A (IgA). Binding of an antigen to a B cell receptor initiates a humoral immune response, binding to a T cell receptor initiates a cellular immune response.

When an animal encounters an antigen for the first time it will have only a few antigen-specific lymphocytes in the adaptive immune system. These few cells must then undergo clonal expansion and differentiation into active effector cells before the pathogen can be destroyed. Consequently the adaptive immune system is slow in responding to the first exposure of the pathogen and the response can take up to four days. However unlike the more rapid acting innate system, the adaptive system leaves behind a pool of previously activated lymphocytes and a "memory" of preformed antibodies that can act immediately upon subsequent infection. Animals that have an appropriate number of activated specific T cells in their blood are immune to a particular pathogen. This is widely exploited in vaccination programmes.

The essence of the immune system is its capacity to recognise surface features of macromolecules that are not normal constituents of the body of the animal. This specific recognition is mediated by serum proteins or antibodies and by antibodies on the surface of T-lymphocytes. Antibodies belong to a class of proteins called immunoglobulins and can be divided into; IgM, IgG, IgA, IgD, and IgE. IgA antibodies act as a protective barrier against micro-organisms in the gastrointestinal tract. IgA is also a major immunoglobulin in milk and colostrum where it may function to protect the gastrointestinal tract of young animals.

Antibodies perform two distinct functions: antigen recognition and elimination. The basic structures and gross chemical properties of immunoglobulins are very similar but their combining specificities vary widely, reflecting the spectrum of antigens that the animal has encountered during its lifetime. Antibody molecules can interact with a virtually unlimited number of antigens.

Many infections are accompanied by an acute phase response which is characterised by synthesis of hepatic acute phase proteins. In mammals and poultry cytokines produced by macrophages during an immune response induce the hepatic secretion of a variety of acute phase proteins which are produced earlier than specific antibodies. This also leads to development of fever, and accelerated protein turnover in the body. Acute phase proteins help the innate and adaptive immune systems. The acute phase response includes changes in metabolism in all organ systems and especially in the liver and muscles. The acute phase response is likely to consume significant amounts of nutrients during an infection and this needs to be taken account of in Total Nutrition.

Immunopathology

The immune system is extremely complicated. It is comprised of many different cells such as lymphocytes and macrophages together with specific molecules, including innumerable antibodies and cytokines. Consequently it is not surprising that the system can malfunction and cause a range of health problems for the animal.

Immunopathology or diseases of the immune system can be grouped into two classes: Deficiency diseases or immunosuppression, when a component of the system fails to function. These diseases are manifested by low resistance to infection. The other problem is hypersensitivity or loss of tolerance when the system reacts under inappropriate conditions. These various immune problems lead to a variety of pathological symptoms. Antigenic stimulation causes several categories of cells to proliferate and differentiate and impair growth rates. Uncontrolled continuation of an immune response leads to pathological lymphoid proliferation, excessive antibody production and even anti-self reactions where the immune system attacks cells of the host.

Immunosuppression

Immunosuppresssion is a recurring economic problem in commercial pig and poultry production. It may be caused by a number of factors such as stress, nutrition, viruses, bacteria and mycotoxins. This may appear as increased susceptibility to specific diseases or as sub-clinical infections and poor response to vaccines. A major problem worldwide for pig production has been the emergence of immunosuppressive viruses such as porcine circovirus 2, porcine reproductive and respiratory disease (PRRS) and swine influenza. The rapid spread of these viruses predisposes animals to secondary infections and has brought forward acute respiratory disease from the finishing stage to the post-weaning stage. Some of the more important viral syndromes caused by immunosuppressive viruses occurring in pigs and poultry are listed in Table 2.

Mycotoxins are produced by moulds that contaminate feeds and are well known immunosuppressive agents. Good standards of feed hygiene are important to minimise mycotoxin production as discussed in Chapter 2.

Immunosuppresed animals or poultry will be more susceptible to any opportunistic viral or bacterial infection. In poultry a general immunosuppression also results in poor immune response of many birds to Newcastle disease vaccine, and to an increase in infection of the flock by *E. coli*.

Animal species	Virus or disease syndrome
Pigs	Porcine circovirus 2 Porcine reproductive and respiratory disease (PRRS) Swine Influenza
Poultry	Marek's disease Infectious bursal disease (Gumboro) Reticuloendotheliosis Chicken infectious anaemia Reovirus (viral arthritis/tenosynovitis) Haemorrhagic enteric virus (turkeys)

Table 2
Immunosuppressive viral syndromes infecting pigs and poultry

In dairy cows several factors may compromise immune mechanisms such as an inflammatory response in the udder elicited by bacteria infecting cows in early lactation, the physiological stress of lactation, ketosis, and milk fever. After parturition there maybe a decrease in the number of circulating neutrophils in the blood capable of phagocytosis and of killing bacteria.

Immunodeficiency may also be a secondary consequence of diseases which result in the accumulation of immunosuppressive products. For example, poor functioning of the kidney or liver leads to an accumulation of toxic substances, which may depress immune responses. Virus infections are often associated with the release of immunosuppressive products into the blood.

Immune tolerance

The immune system has to process many harmless antigens and must be able to discriminate between antigens with the potential to cause damage and those without. This is the phenomenon of immune tolerance and is a mechanism that prevents organisms from mounting immune responses to the antigenic determinants of their own macromolecules and to harmless feed molecules. Feed-derived antigens and bacterial antigens from the gut microflora trigger immune tolerance in healthy animals whereas pathogens induce strong activation of immune defence mechanisms required to control pathogens. Some feed proteins, for example soy proteins, are known to generate unnecessary immune response in young animals. In these cases immune tolerance is not as effective as it could be. Failure of tolerance leads to digestive diseases such as allergic reactions and inflammatory problems in the large intestine. Immune tolerance is extremely important but is still poorly understood and difficult to manipulate.

The mucosal immune system probably generates tolerance, rather than active responses to the majority of antigens to which it is exposed. This is known as the regulatory function. At the same time, it must retain the ability to respond actively to potential pathogens. This is known as the effector function. The immune system in an animal must continually balance these effector and regulatory functions in response to the antigens which it encounters and then make the appropriate immune response. The wrong decision will result in a loss of intestinal function as a result of inadequate control of pathogens or an allergic response to feed.

The immune system has evolved to react aggressively against foreign molecules. However inevitably small quantities of harmless but antigenic proteins are absorbed from the intestine of humans, rodents and pigs (Bailey *et al.*, 2001). Whilst nutritionally insignificant these proteins are enough to trigger strong, damaging, allergic immune responses in the intestine. Normally these responses are actively prevented by the phenomenon of immune tolerance. Efficient and accurate discrimination between harmful and harmless antigens and the expression of appropriate responses to each is a manifestation of good health in an animal.

An efficient, decision-making mucosal immune system is a particular requirement in young animals such as neonatal piglets and newly hatched chicks. After birth and after weaning, the young piglet will be exposed to a huge array of novel antigens associated with feed and with pathogenic micro-organisms. However, in addition to the absence of endogenous antibody, young piglets have essentially no mucosal immune system and are unable to initiate an appropriate response to either a dietary or microbial challenge. This system develops after birth approaching maturity after five weeks. The system develops as a balance between effector and regulatory function and during weaning this balance is disturbed and the weaned piglet is predisposed to bacterial infection and disease.

Newly hatched chicks are similarly highly susceptible to infection by opportunistic pathogens during the first two weeks of life. Several factors contribute to this problem, including immaturity of the immune system, declining maternal antibody protection and stress factors associated with intensive raising of poultry.

It appears that the developing flora in the gastrointestinal tract is critical in establishing normal immune function. It is likely that disruption of the microflora and the immune system at point of weaning may have long lasting effects on mucosal function. The example, of non-specific colitis in growing pigs may reflect a chronic inability to respond to mucosal antigens, and may be a consequence of weaning practice.

If development of the microflora in the gastrointestinal tract is essential for maturation of the immune system then addition of selected micro-organisms as probiotics might contribute to its development. Probiotics have been widely studied and are quite extensively used but more for a competitive exclusion function than for stimulation of the immune system (Chapter 5). They may have some beneficial effects in enhancing the innate immune system, particularly increased macrophage activity (McCracken and Gaskins, 1999).

Immune activation

The presence of bacterial and viral pathogens in feed or in the environment cause leukocytes to secrete cytokines which are proteins found throughout the body to activate both the cellular and humoral components of the immune system. The cytokines are a complex group of molecules produced by the body and include the interleukines and tumour necrosis factors. They bring about changes designed to deliver nutrients to the immune system. Cytokine release leads to rapid proliferation of lymphocytes and macrophages, antibody production, and hepatic acute phase protein synthesis. The activated immune system then assists in the repair of tissues, protects healthy tissue from the effects of free radicals and other oxidising molecules and removes nutrients from the bloodstream which might assist multiplication of the pathogen.

Cytokines are also important messengers used by the immune system to inform the rest of the body of a challenge. They generate similar metabolic responses in both poultry and pigs; reduced feed intake, increased body temperature and reduced growth rates (Johnson, 1997, Klasing *et al.*, 1991, Williams *et al.*, 1997a and 1997b). The cytokine signal from the immune system allows the body to re-direct its metabolic activities and diverts nutrients away from growth towards fighting the disease challenge. This reduction in growth due to activation of the immune system is now seen as a major obstacle to growth efficiency and is important in the raising of animals for food where maximum growth is required. Diets which do not activate the immune system unnecessarily will be more advantageous in terms of the efficiency of animal production.

Rapid animal growth compounded with immune system activation results in both an increase and a decrease in nutrient requirements. The increase in body temperature and fever associated with cytokine production increases energy demands. Fever is associated with an increase in basal metabolic rate of 10-15% for each 1°C above normal. There is greatly increased proliferation of cells of the immune system, lymphocytes and macrophages, antibody production and synthesis of hepatic acute phase proteins is increased. This demands extra energy supplies. There is however also a reduction in energy demands due to reduced activity and increased sleep and reduction in the growth rate.

In growing pigs the amino acid requirements are increased due to breakdown of proteins for use by the immune system or to use as an energy source in gluconeogenic reactions. However there is also a lower requirement for amino acids due to reduced protein accretion and slower growth.

Pigs raised under either high or low levels of antigen exposure which in turn gave either high or low levels of immune activation had quite different growth performances (Williams *et al.*, 1997a and 1997b). Piglets from 6 to 27 kg with low immune system activation had bigger appetites, faster growth and superior feed conversion efficiency compared to those piglets with a high activation of the immune system. There were also significant differences in protein gain and body leanness. Piglets with a high activation of the immune system had a much reduced lysine intake compared to piglets with a low activation of the immune system (Table 3).

Table 3
Effect of activation of the immune system on growth performance of piglets from 6-27 kg (lysine 1.2%)

Growth characteristic	Immune system activation	
	High (poor health status)	Low (good health status)
Daily feed intake (g)	889	1052
Daily lysine intake (g)	8.78	12.21
Average daily gain (g)	510	644
FCR	1.72	1.63

These performance differences persisted throughout the life of the pigs up to 112 kg. Low antigen exposure resulted in an increased appetite, higher growth rates, better leanness, and improved feed conversion efficiency compared to pigs raised with high antigen exposure. The low immune status pigs reached market weight 20 days earlier and with 42 Kg less feed.

Detailed studies of varying lysine levels indicated that the daily intake and dietary concentration of lysine needed to give optimum performance depended on the level of immune activation of the pigs. The higher feed intake of the pigs with low immune activation also required extra lysine needed for optimum growth performance. Growth rate and feed conversion efficiency were optimal in pigs with low immune system activation with lysine at 1.5% of the feed. Pigs with high activation of the immune system only required lysine at 1.2% of the feed. This is probably because immune system activation shifts the balance between growth and maintenance. Lysine is present in lean tissue growth at about three times the level needed for proteins associated with maintenance. This clearly indicates that immune status has an important influence upon nutrient utilization and would suggest that diets should be formulated taking immune status into account.

This is made all the more complicated by the unpredictable degree of appetite reduction in animals with high activation of the immune system and by the difficulties in actually measuring immune status in animals. This is a far from trivial task and will be further discussed in Chapter 8 on assessment of Total Nutrition. Nevertheless it is a task worth pursuing as there are serious economic penalties if feed formulations do not match the immune status of the animals. When animals are raised under a high disease challenge they need to invest more nutrients in supporting the immune system and this results in slower growth and less yield of meat per tonne of feed consumed.

The performance of pigs under a high disease challenge versus a low disease challenge in terms of growth rate and feed efficiency can be quite significant up to 10-25%. This could translate into a growth rate reduction of up to 150 g/day over 6-92 kg. This means about four weeks longer growth period to reach slaughter weight for the pigs with the high disease challenge. Pigs raised under both high and low disease challenge can appear equally healthy but have different growth rates. Pigs with a high disease challenge have a highly activated immune system and in defending against diseases devote substantial amounts of nutrients to combating the disease challenge in order to stay healthy. By contrast pigs with a low disease challenge and a low activation of the immune system can use more of their dietary nutrients for growth as they do not have to support an activated immune system. Consequently these two situations require different feed formulations.

There is a marked difference in lysine and in other amino acid requirements for pigs with a high or low disease challenge (Williams *et al.*, 1997a, 1997b). The economic penalties arise from under-specifying diets for pigs with low disease challenge and over-specifying diets for pigs with high disease challenge. In pigs with high disease challenge energy is used to excrete amino acids which cannot be utilised in tissue growth so a high specification diet is wasted in this situation.

Nutritional immunology (immunonutrition)

Good health and performance of food animals depends upon many factors including genetics, housing, the frequency of exposure to pathogens and vaccination programmes. Nutrition however also plays a major role in modulating the susceptibility of an animal to infectious diseases. Nutrition and in particular protein supply is important for the development of resistance in many animals including lambs and mature sheep (Donaldson *et al.*, 1998). This may reflect the very high ratio of requirement for amino acid relative to metabolizable energy of the young animal and the possibility that infection by pathogenic organisms increases protein requirement. Milk proteins, both casein and a whey mixture, induced spleen cells to increase production of immunoglobulin M (IgM), (Wong *et al.*, 1998). This may explain some of the benefits frequently seen from incorporation of whey products into feed for young animals such as piglets and calves.

It is well established that nutritional deficiency will be associated with an impaired immune response to disease challenges. Nutritional deficiency affects cell-mediated immunity, antibody production and cytokine production and leads to a general immunodeficiency. The capacity for nutrients to modulate the actions of the immune system is an important issue in modern animal production.

The application of nutrients to modulate the immune system is known as nutritional immunology or immunonutrition. This may be defined as the modulation of the activities of the immune system by specific nutrients or nutricines fed in amounts above those normally encountered in the diet (Grimble, 2001). Nutritional immunology is an attractive method to regulate the growth inhibiting effects of immune activation. However, control of immune system activation is very difficult for animal nutritionists but nevertheless Total Nutrition must make use of nutritional immunology whenever possible.

Knowledge of the amounts and types of essential nutrients is well established. Modern feed formulations and nutritional strategies usually supply adequate amounts of basic nutrients and avoids overt problems of nutrient deficiencies. This may not be the same however as supplying an adequate level of nutrients and nutricines to maximize productivity in healthy animals and also to maintain immunocompetence with good disease resistance. Considerable work will be needed within the concept of Total Nutrition to develop a thorough understanding of the mechanisms by which nutrition influences both growth and the immune system.

Feed is a source of nutrients for the initial development of the immune cells and various effector molecules and during an actual immune response. However, feed is also the largest source of antigens (and other chemicals) encountered by the body, and it may contain components such as mycotoxins, capable of modifying the immune system.

The presence of antigenic materials and of pathogens in feed results in the activation of the immune system and in a rapid release of a wide range of potent molecules which include cytokines, nitric oxide, hydrogen peroxide and superoxide and hydroxyl and hypochlorite free radicals. Nitric oxide, hydrogen peroxide and free radicals provide a potent defence by means of a chemical attack upon the membranes, cellular proteins and nuclear materials of the invading pathogenic organism.

The components of the immune system are strongly influenced by nutrition and further knowledge in this area is important. Nutrition will enhance the ability of the immune system to respond to a disease challenge. It will influence the passive transfer of immune protection in milk or through eggs to the progeny.

At present there are a relatively limited number of nutrients and nutricines which are recognised to have an important function in nutritional immunology (Table 4).

Table 4
Feed components important in immunomodulation

Immunomodulator	Function
Arginine	Substrate for nitric oxide (NO) synthesis, improves helper T-cell numbers.
Carotenoids	Antioxidant function, stimulates vaccine response.
Cysteine	Enhances antioxidant status via glutathione synthesis
Flavonoids	Enhances virus elimination from blood
Glutamine	Nutrient for immune cells, improves gut wall functions, precursor for glutathione.
Nucleotides	RNA and DNA precursors, improves T-cell function
n-3 polyunsaturated fatty acids	Anti-inflammatory agents, reverses immunosuppression.
Zinc	Maintains T-cell response and antibody production

Several amino acids such as arginine, cysteine and glutamine play important roles in supporting the immune system. These amino acids may need to be considered as conditionally essential nutrients (Lacey and Wilmore, 1990). The original concept of defining nutrient requirements on the basis of deficiency levels is not relevant in Total Nutrition as many feed ingredients considered here will not have classic deficiency levels. Conditionally essential nutrient however is a more useful concept because it encompasses nutrients that may be synthesised by the body and are therefore in the classic sense non-essential. However endogenous synthesis of these nutrients may be inadequate to meet the requirements of the animal under various conditions such as immune stress. Consequently Total Nutrition must look at both nutricines and conditionally essential nutrients in order to formulate diets for good health as well as growth.

Arginine may be particularly important in poultry since unlike most mammals they cannot synthesize arginine and thus it is an essential amino acid in poultry nutrition. In addition to its function in protein synthesis, arginine seems to have beneficial effects upon several aspects of the immune system and influences disease resistance. Arginine supplementation of poultry feed markedly influenced development of the immune system organs with a pronounced effect upon the thymus and spleen (Kwak, *et al.*, 1999).

Cysteine is a limiting substrate for the biosynthesis of glutathione which in turn is a limiting factor for the immune system and is an important cellular antioxidant (Chapter 7). The cysteine supply and the intracellular glutathione levels have a strong influence on the T-cell system because these cells have a particularly strong demand for cysteine during the rapid increase of cell volume after antigenic stimulation (Dröge *et al.*, 1994).

Glutamine is used by cells of the immune system as an energy source. Any limitations in glutamine supply would decrease the rate of proliferation of cells of the immune system and decrease their ability to respond rapidly to immune challenges. Glutamine is a critical nutrient for the maintenance of the immune system in the gastrointestinal tract. Supplementation of feeds with glutamine was quite beneficial in influencing the immune system in piglets infected with *E. coli* (Yoo et *al.*, 1997) and a similar response was seen with arginine (Flynn and Wu, 1997).

In sick animals a conditional deficiency of glutamine could occur if increased metabolic demands exceed the endogenous supply. Therefore the glutamine content of a feed may become more critical when animals are under severe infectious pressure or become sick. In critically ill humans glutamine is now seriously considered as a valuable component in supporting recovery of such patients (Griffiths, 2001) and this may become more widely utilized in animal nutrition. There is a case for more careful consideration of the glutamine contents of animal diets perhaps.

Various antioxidants, in particular the carotenoids such as lutein, are valuable in supporting the immune system. Many animals can rapidly absorb lutein from the feed and some of this absorbed lutein is deposited in the spleen (Park *et al.*, 1998). This suggests a role for lutein in modulating immunity. This has been further confirmed in a study with dietary lutein in dogs which showed a stimulation of both cell-mediated and humoral immune responses (Kim *et al.*, 2000). In this work dogs were fed lutein in the diet for up to 17 weeks. Lutein did not significantly influence the concentrations of plasma IgM or IgG during the first 12 weeks of the study. After an initial 12 weeks period dogs were then inoculated with a polyvalent vaccine on weeks 13 and 15. After week 15 plasma IgG concentrations increased significantly in lutein fed dogs (Table 5). This suggests that dietary lutein may enhance antibody response of dogs given routine vaccinations. Carotenoids are already recognised as playing an important role in non-infectious diseases (Chapter 7) but they also have an important function in regulation of the immune system.

Table 5

Changes in plasma IgG concentration (mg/ml) in dogs fed lutein

Lutein	Weeks				
(mg/day)	13 (Vacc.)*	14	15 (Vacc.)	16	17
0	12.7	13.0	12.7	12.8	13.1
20	13.1	13.8	13.6	14.0	15.0

Note: * polyvalent vaccine administered to all dogs

Fish oils and conjugated linoleic acid (a structural isomer of linoleic acid) have been reported to be effective in preventing weight loss following immune stimulation (Cook *et al.*, 1993). The beneficial effects of fish oil in supporting the immune system is widely accepted and probably is related to the intake of the long-chain polyunsaturated fatty acids in the fish oils (n-3 PUFA). Polyunsaturated fatty acids are an important component of pig diets and may play a valuable role in helping immune system development and immune responsiveness. After an extensive review of PUFA in pig development it was concluded that pig feed formulations should supply 8% of the dietary energy as linoleic acid (Leskanich and Noble, 1999).

A group of flavonoids that are found predominantly in soyabeans and clover, isoflavones, are an interesting example of feed components with potentially useful immuno-modulating effects. One particular isoflavone found in soyabean meal is genistein and at 200-400 g/tonne of feed enhances elimination of virus from the serum of pigs and improves growth performance in pigs suffering a virus challenge (Greiner *et al.*, 2001). Furthermore a reduction in virus concentration in blood serum was observed when genistein was fed to the pigs although the virus was not completely eliminated. Nevertheless it seems that reduction in virus titre is able to improve growth performance of the animals and this offers the exciting possibility of use of feed ingredients against viral diseases.

Zinc is widely recognised as a micronutrient that is essential for optimum immune response. Zinc deficiency leads to reduced production of macrophages and neutrophills, reduced T-cell responses and reduced antibody production after a microbial challenge. Zinc affects multiple aspects of the immune system and adequate zinc levels are crucial nutrient for normal development of the immune system (Shankar and Prasad, 1998).

Nutritional immunology is a complex subject but one which must be more widely exploited for the successful raising of animals for food where ever less reliance upon drugs and medicines is required. As illustrated in Figure 2 various compounds play an important role as immunonutrients and it will be important to ensure that feeds supply appropriate amounts of these.

Activation of the immune system and production of inflammatory cytokines is associated with a reduction in growth rate and loss of body tissue. Nutrients are released from tissues to support the synthesis of glutathione, acute phase proteins and the production of T- and B-cells. This will strengthen the antioxidant defences and destroy pathogens. It is also clearly important that the environment and the feed delivers the minimum amount of infectious pathogens and antigenic feed components to the animal.

Figure 2
Central role of
nutritional immunology
in maintenance of
animal health

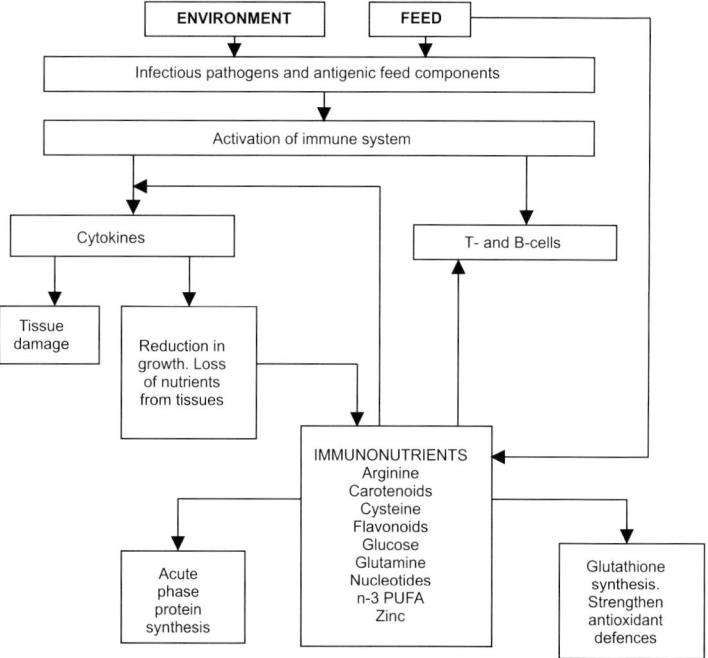

Formulating appropriate feeds is quite complex because immune stimulation results in decreased feed intake as well as in poorer feed conversions or decreased efficiency in use of nutrients. Reduced growth rates from decreased feed intake or reduced efficiency have different economic costs. Inefficient use of ingested nutrients for growth will have a larger economic impact than decreased nutrient intake. This is because reduced feed conversions incurs the additional expense of increased nutrients used per kilo of gain. Although decreased nutrient intake also results in decreased gain it does not incur increased costs per unit of gain. The major economic loss here is that fewer kilos of meat can be produced from a unit in terms of time and space.

There are evidently many important and complex interactions between animal nutrition, the activity of the various defence systems and growth and performance. Understanding these interactions is an important part of Total Nutrition and will enable the nutritionist to contribute to the economics of food production, animal welfare and to protect consumers from food-borne pathogens.

Future directions

Animals have an enormous potential to convert feed ingredients rapidly and efficiently into lean tissue when the effect of various antigens is limited. This will require nutritionists in the future assess the immune status or disease challenge of the animals in each location and formulate feeds for rapid growth, high productivity and disease management.

Disease management

A general objective must be to develop knowledge and systems to enhance the general defence systems of animals to better withstand infectious diseases. This is increasingly important for reasons of both economics and animal welfare. The objections to the use of antibiotics and other drugs in animals raised for food make it imperative that the inherent defence systems of the animals are strengthened and supported. It will be difficult to affect changes in such a complex biological system with so many different components comprising both effector molecules and specific cells. Nevertheless Total Nutrition must in the future consider feeding to improve resistance of the animal to infectious diseases as well as to improve growth and productivity.

There are several characteristics of the defence system which need to be addressed as listed in Table 6. Regulation of the immune system is very complex and these objectives will not be achieved easily or quickly. Current work is directed towards correction of evident problems caused by weaning stress, immunosuppression or excessive activation of the immune system. There is continuing activity in vaccines development as manifested by recent developments in vaccines against *Salmonella* and coccidiosis.

Table 6
Objectives to
improve defence
responses of
animals to
infectious
diseases

Novel peptides
Accelerate development of immune response in young animals
Immunostimulation to extend or sustain immune response
Mitigate immunosuppression arising from mycotoxins, stress or other
    diseases
Improve tolerance to non-pathogenic environmental and feed antigens
Mitigate growth inhibiting effect of the immune system
Immunotherapy, novel techniques of disease control
Enhance vaccine response

Peptides

Peptides may have some practical significance in the future as pathogen control agents. They are attractive in that they have a wide spectrum of antimicrobial activity, low mammalian toxicity and are quite heat stable. This would make them ideal candidates for inclusion into animal feeds. However at the present only the peptide nisin is in commercial use as an antimicrobial agent and this is used directly in human food products such as dairy products, milk and canned foods (Henning, *et al.*, 1986). Nisin is produced by fermentation and it is far too expensive for use in animal nutrition. A possible solution is the chemical synthesis of bio-active peptides. A peptide based only on the amino acids leucine and lysine had non-specific activity against Gram-negative and Gram-positive bacteria and against yeasts (Appendini and Hotchkiss, 2000). It should be possible to synthesize a range of peptides with different antimicrobial activity in a cost-effective manner. These could become important feed ingredients in Total Nutrition.

Immunomodulation

Immunomodulation is a difficult concept in scientific terms because there is the necessity to balance immunostimulation against excessive activation of the immune system which is growth-inhibiting. Immunostimulating agents are also likely to exhibit an opposite or immunosuppressive effect when increasing doses are applied. Clearly a more robust, rapid acting and sustained immune response would be desirable. Pigs selected for high antibody and cell-mediated immune responses had better rates of gain than pigs with low immune response (Wilkie and Mallard, 1999).

Immunostimulation primarily implies stimulation of the innate or non-specific immune which comprises macrophages, neutrophils, natural killer cells, complement factors and certain T-lymphocyte populations.

Ideally immunostimulants should induce production of various effector substances that not only attack micro-organisms but also play an important role in the regulation of the immune system and hence potentiate or suppress other immunological systems.

It is far from clear as to which molecules and products might be immunostimulatory. Large molecular weight compounds such as polysaccharides, proteins, glycopepetides and nucleotides have been identified as having immunostimulatory properties. ß-glucans are already used in aquaculture as immunostimulants. Addition of ß-glucans to pig feeds did not influence macrophage or neutrophil function but did improve post-weaning growth compared to pigs on a control diet (Dritz *et al.*, 1995). Lactic acid bacteria and fermented milk products have long been thought to have beneficial effects upon immune status. An immunomodulator based on parapoxvirus ORF with various chemical components was effective in limiting the spread of infectious bovine rhinotracheitis in calves (Castrucci *et al.*, 2000).

There is great interest today in the application of various botanical or herbal products as immunomodulators. It is well established that extracts from herbs and spices such as rosemary, sage, thyme and oregano have antioxidant and antimicrobial activities. These could indirectly support the immune system. Some species of *Echinacea* are already utilized as in human nutrition as an immunostimulator (Briskin, 2000). A large number of pharmacological studies have provided strong evidence for the immunomodulating effects of *Echinacea* species. *Echinacea* as a natural product will contain an extremely complex mixture of different molecules and the immunostimulatory effect appears to result from the combined effects of; isobutylamides, caffeic acid derivatives and non-starch polysaccharides. This makes it very difficult to obtain preparations with standardized ingredients and with standardized immunostimulatory activity. Development of suitable assay systems for immunostimulation would be of great advantage here. Certainly there will be increased emphasis in the future on developing products or compounds with immunostimulatory activity and these will most likely come from plants.

Immunotherapy

In the light of the problems engendered by use of antibiotics alternative approaches to disease therapy are evolving and one exciting approach is to extend the various immune system processes. One aspect focuses on carbohydrates that are the key to bacterial binding in the gut. Specific proteins called adhesins present on the surface of bacteria, fungi, viruses and parasites interact with carbohydrate chains on the walls of the cells of the gastrointestinal tract. The sugars mannose and galactose are important here. There are age-related changes in bacterial binding in the gastrointestinal tract and neonates have receptors for binding many pathogens so if they get exposure to these pathogens they will contract a disease. If the animal does not have the specific carbohydrate receptors they will not be sensitive to pathogenic bacteria.

There is the exciting possibility to develop receptor analogues sometimes described as "chemical prebiotics." This requires the production of soluble receptor analogues to bind pathogens in the gut and prevent them causing disease. This is also one of the postulated modes of action of non-digestible oligosaccharides (NDO) as discussed in Chapter 5. The NDO will preferentially bind cells of the pathogen and prevent them from adhering to the gut wall. Lectins are proteins that occur in many plants and are often considered as antinutritional factors in various legumes. However they have very specific carbohydrate recognising structures similar to bacterial adhesins and possibly they could be incorporated into a disease management system.

Another strategy is to immunize breeding poultry against a growth-inhibiting factor such as urease. Many bacteria in the gastrointestinal tract produce urease enzymes that break down urea to carbon dioxide and ammonia. Accumulation of ammonia in the gut and release of ammonia from manure is undesirable. It is always a potential loss of dietary nitrogen and the generation of ammonia in the gastrointestinal tract may be toxic and this could limit growth. Injection of turkey and broiler breeders with urease from the Jackbean plant stimulates production of antibodies to the urease enzyme. This gave some modest benefits in the progeny from the inoculated turkeys and chickens (Pimentel *et al.*, 1991).

Cytokines as described above have an important influence upon the health and immune status of animals. Although they have been generally considered as growth inhibiting agents they could also be valuable naturally occurring therapeutic agents. Cytokines regulate a wide variety of functions such as disease resistance, wound healing, bone development, nutrient partitioning, appetite, growth and reproduction.

Interferons (IFNs) for example, are a family of cytokines that share the capacity to inhibit viral replication and to modulate immune function. Broiler chickens experimentally injected with IFN-γ displayed enhanced weight gain and were less effected by coccidiosis than control birds (Lowenthal *et al.*, 1999). The precise mechanisms of the action of these cytokines is not known but is possibly due to IFN-γ enhancing the immune system. This programme is not applicable in practice in modern poultry production because the rapid degradation and clearance of injected cytokines would require multiple injections. Therefore delivery systems other than by injection need to be developed. Nevertheless it is a good example of a new approach to disease control in poultry that may become significant in the future.

Another application of immunotherapy is possibly in immunocastration (Jago *et al.*, 1999). Active immunization against gonadotrophin-releasing hormone has been investigated as an alternative to traditional methods of castrating bulls. However initial studies suggest that immunocastration is effective but is likely to result in reduced weight gain of the animals.

Antioxidants and immune function

Control of oxidative stress is important in avoidance of non-infectious diseases as discussed in Chapter 7. However oxidative damage is a particular hazard of the immune system because many immune cells produce reactive oxygen species as part of the defence against infection. Therefore it will be important to ensure that adequate amounts of antioxidants are formulated into diets to protect tissues against oxidative damage as part of the immune response.

There is also the possibility that the nutritional status of an animal can have a direct effect upon the virulence of some viruses. A normally-benign strain of coxsackievirus B3 will become virulent in selenium or vitamin E deficient mice (Beck, 1999). This change in virulence is due to specific mutations in the virus itself such that once the mutation has occurred even mice with normal nutrition become vulnerable to the virus. A similar phenomenon has been observed in pigs with porcine circovirus 1 which is non-pathogenic but which has mutated into porcine circovirus 2 which is highly pathogenic.

Studies with coxsackievirus in mice indicated that oxidative stress in the host animals leads to mutations in the virus which in turn leads to increased virulence of the particular virus. A virulent stain of coxsckievirus became even more virulent under oxidative stress. These observations are of enormous importance for future developments in Total Nutrition. If oxidative stress on an animal can cause changes in a viral pathogen then the host nutritional status will play an important role in the virulence of existing pathogens and in the emergence of new viral pathogens. Now a nutritional stress affects not only the host but may also directly increase virulence of pathogenic viruses. This phenomenon has enormous implications for animal health and control of viruses.

Antioxidants are likely to become ever more important; to maintain an efficient immune status, to avoid mutation of viruses into more virulent pathogens, and to avoid non-infectious diseases. Much further work needs to be done to establish this link between nutrition of animals and mutation of viruses and to investigate appropriate dietary antioxidant nutricines.

Growth regulation

The immune system is primarily considered as a defence system to protect the animal against threats from the environment. However it is also possible to generate antibodies in the immune system which will either neutralize the activity of growth inhibiting factors or potentiate the activity of growth stimulating factors. Antibodies could be either produced in one animal and administered to a second animal or directly induced in the animal of concern by immunization with specifically designed antigens.

Antibodies can be used as hormone mimics and growth hormone is a good candidate for this approach. Growth hormone is essential for normal growth of animals and exogenous growth hormone can improve the deposition of lean tissue. Somatostatin inhibits growth hormone secretion. Therefore immuno-neutralization of somatostatin should stimulate production of growth hormone resulting in increased growth hormone concentrations in the blood and enhanced growth of the animal. Results have been quite variable and the technique is still not fully developed (Pell, 1997).

Vaccines

Immunization of animals has over the years proven to be extremely effective in preventing economic losses and animal suffering, from many infectious diseases. Vaccines not only reduce the severity of disease in vaccinated animals but they also reduce the transmission of infectious agents from vaccinated animals that accidentally become infected with the agent. This results in a reduced loss of production in all contact animals. In general live attenuated or killed conventional vaccines are used.

Vaccines have their limitations in that most are effective against viruses, there are fewer vaccines available for bacteria and possibly only against parasites,for coccidiosis. Vaccines are more widely developed to control respiratory diseases and relatively few are available for enteric diseases. The elimination of antibiotic growth promoters in poultry will bring increased incidences of necrotic enteritis caused by the Gram-positive organism *Clostridia perfringens* and a vaccine would be very useful to control this problem.

New developments in biotechnology have led to the concept of vectored vaccines. These are live, non-pathogenic organisms used to deliver a foreign antigen to the cells of the immune system. The advantage of such live vaccines is that they invoke a strong immune response with a single dose and that the effect is long-lasting. Also there is the possibility of designing multi-valent antigen systems to give simultaneous protection against a range of pathogens using a single dose of live vaccine. Vectored vaccines could deliver bacterial, viral and parasitic antigens.

The use of vectored vaccines would overcome the problems associated with modified (attenuated) live vaccines in the past. There was always the risk of a live vaccine strain reverting to full virulence in the animal and causing an outbreak of the disease supposedly being controlled. Now absolute safety can be built into live vaccine products by using as vectors bacterial strains which cannot cause disease due to some genetic modification of the organism.

Another strategy in vaccine development is DNA vaccines since naked DNA is a very effective means of eliciting an immune response to the antigens encoded in it. In principle DNA vaccines should be simple to produce and they will not require purification of proteins or organisms as do conventional vaccines. A useful property of DNA vaccines is their ability to elicit both cytotoxic and humoral (antibody) responses and an important advantage is their safety. Since the vaccines consist of a single gene, there is no possibility to induce an infection or subclinical disease by vaccination. Many DNA vaccines do not produce very high levels of antibody, but they are extremely efficient primers of the immune system. On subsequent exposure to a pathogen there is a rapid immune response which rapidly clears the infectious agent. DNA vaccines could also be used to immunize animals at birth. This is not possible with conventional vaccines because neonates usually do not respond, as they possess some maternally derived antibodies. Unfortunately the level of maternal antibody which inhibits the response to conventional vaccines is often lower than that required for protection. DNA vaccines are not inhibited by the presence of maternal antibodies. DNA vaccines have already been developed and tested in a wide range of animal species including, cattle, sheep, horses, pigs, dogs, cats and fish (Babiuk *et al.*, 1999).

There are also concerns about DNA viruses which need to be explored more fully. The main issue is the slight chance of random integration of the DNA vaccine into the genome of the host cells. Another unresolved issue is the persistence of DNA which may remain active for several months after immunization. This could conceivably develop tolerance to the antigen and reduced protection. This would be more significant in treatment of humans, as animals raised for food normally have relatively short life cycles.

Another strategy to protect young animals is to feed them specific antibodies produced in egg yolks. This system takes advantage of the fact that chickens do not become tolerant to antigens on repeated exposure but continue to produce antibodies to a specific antigen (Miller and Cook, 1994). By contrast pigs, dogs or humans readily become tolerant to antigens upon repeated exposure. Furthermore the laying hen will also deposit antibodies in the egg yolk. Consequently specific antibodies can be induced in egg yolk and the spray-dried egg fed to other species as a source of antibodies. Some preliminary results with piglets have shown improved growth performances and this process may become more valuable in the light of the reduction in use of antibiotics (Kichura, 1998).

References

Appendini, P. and Hotchkiss, J. H. (2000). Antimicrobial activity of a 14-residue synthetic peptide against foodborne microorganisms. *Journal of Food Protection*, **63**, 889-893.

Babiuk, L. A., van Drunen, S., den Hurk, L-v. and Babiuk, S. L. (1999). Immunization of animals: from DNA to the dinner plate. *Veterinary Immunology and Immunopathology*, **72**, 189-202.

Bailey, M., Haverson, K., Vega-Lopez, M. A., Bland, P. W., Miller, B. G. and Stokes, C. R. (2000) Enteric immunity and gut health. In: *The Weaner Pig*. Eds: M. A. Varley and J. Wiseman. CABI Publishing, Wallingford, UK. pp. 207-223.

Beck, M. A. (1999). Selenium and host defence towards viruses. *Proceedings of the Nutrition Society*, **58**, 707-711.

Briskin, D. P. (2000). Medicinal plants and phytomedicines. Linking plant biochemistry and physiology to human health. *Plant Phsyiology*, **124**, 505-514.

Castrucci, G., Osburn, B. I., Figeri, F., Ferrari, M., Salvatori, D., Lo dico, M and Barreca, F. (2000). Then use of immunomodulators in the control of infectious bovine rhinotracheitis. *Comparative Immunology, Microbiology and Infectious Diseases*, **23**, 163-173.

Conley, A. and Kabara, J.J. (1973). Antimicrobial action of esters of polyhydric alcohols. *Antimicrobial Agents and Chemotherapy*, **4**, 501-506.

Cook, M. E., Miller, C. C., Park, Y. and Pariza, M. (1993). Immune modulation: Nutritional control of immune-induced growth depression. *Poultry Science*, **72**, 1301-1305.

Donaldson, J., van Houtert, M. F. J. and Sykes, A. R. (1998). The effect of nutrition on the periparturient parasite status of mature ewes. *Animal Science*, **67**, 523-533.

Dritz, S. S., Shi, J., Kielan, T. L., Goodband, R. D., Nelssen, J. L., Tokach, M. D., Chengappa, M. M., Smith, J. E. and Blecha, F. (1995). Influence of dietary ß-glucan on growth performance, non-specific immunity and resistance to Streptococcus suis infection in weanling pigs. *Journal of Animal Science*, **73**, 3341-3350.

Dröge, W., Schulz-Osthoff, K., Mihm, S., Galter, D., Schenk, H., Eck, H-P., Roth, S. and Gmünder, H. (1994). Functions of glutathione and glutathione disulfide in immunology and immunopathology. *FASEB Journal,* **8**, 1131-1138.

Flynn, N. E. and Wu, G. Y. (1997). Glucocorticoids play an important role in mediating the enhanced metabolism of arginine and glutamine in enterocytes of postweaning pigs. *Journal of Nutrition,* **127**, 732-737.

Greiner, L. L., Stahly, T. S. and Stabel, T. J. (2001). The effect of dietary soy genistein on pig growth and viral replication during a viral challenge. *Journal of Animal Science,* **79**, 1272-1279.

Griffiths, R. D. (2001). The evidence for glutamine use in the critically-ill. *Proceedings of the Nutrition Society,* **60**, 403-410.

Grimble, R. F. (2001). Nutritional modulation of immune function. *Proceedings of the Nutrition Society,* **60**, 389-397.

Henning, S., Metz, R. and Hammes, W. P. (1986). New aspects for the application of nisin to food products based on its mode of action. *International Journal of Food Microbiology,* **3**, 135-141.

Jago, J. G., Matthews, L. R., Trigg, T. E., Dobbie, P. and Bass, J. J. (1999). The effect of immunocastration 7 weeks before slaughter on the behaviour, growth and meat quality of post-pubertal bulls. *Animal Science,* **68**, 163-171.

Johnson, R. W. (1997). Inhibition of growth by pro-inflammatory cytokines: an integrated view. *Journal of Animal Science,* **75**, 1244-1255.

Kichura, T. S. (1998). Commercial feed trials evaluating pig performance with specialty egg protein. *Proceedings of American Association of Swine Professionals*, pp. 189-191.

Kim, H. W., Chew, B. P., Wong, T. S., Park, J. S., Weng, B. B. C., Byrne, K. M., Hayek, M.G. and Reinhart, G. A. (2000). Dietary lutein stimulates immune response in the canine. *Veterinary Immunology and Immunopathology*, **74**, 315-327.

Klasing, K. C. and Johnstone, B. J. (1991). Monokines in growth and development. *Poultry Science,* **70**, 1781-1789.

Kwak, H., Austic, R. E. and Dietert, R. R. (199). Influence of dietary arginine concentration on lymphoid organ growth in chickens. *Poultry Science,* **78**, 1536-1541.

Lacey, J. M. and Wilmore, D. W. (1990). Is glutamine a conditionally essential amino acid ? *Nutrition Reviews,* **48**, 297-309.

Leskanich, C. O. and Noble, R. C. (1999). The comparative roles of polyunsaturated fatty acids in pig neonatal development. *British Journal of Nutrition,* **81**, 87-106.

Lowenthal, J. W., O'Neil, T. E., David, A., Strom, G. and Andrew, M. E. (199). Cytokine therapy: a natural alternative for disease control. *Veterinary Immunology and Immunopathology,* **72**, 183-188.

McCracken, V. J. and Gaskins, H. R. (1999). Probiotics and the immune system. In: *Probiotic. A Critical Review*. Ed: G. W. Tannock. Horizon Scientific Press, Norfolk, UK, pp. 85 - 111.

Miller, C. C. and Cook, M. E. (1994). Evidence against the induction of immunological tolerance by feeding antigens to chickens. *Poultry Science*, **73**, 106-112.

Nicolas, P. and Mor, A. (1995). Peptides as weapons against microorganims in the chemical defense system of vertebrates. *Annual Review of Microbiology*, **49**, 227-304.

Park, J. S., Chew, B. P. and Wong, T. S. (1998). Dietary lutein absorption from marigold extract is rapid in BALB/c mice. *Journal of Nutrition*, **128**, 1802-1806.

Partanen, K. H. and Mroz, Z. (1999). Organic acids for performance enhancement in pig diets. *Nutrition Research Reviews*, **12**, 117-145.

Pell, J. M. (1997). Immunological manipulation of growth. *Proceedings of the Nutrition Society*, **56**, 621-630.

Pimentel, J. L., Cook, M. E. and Jonsson, J. M. (1991). Increased growth of chicks and poults obtained from hens injected with Jackbean urease. *Poultry Science*, **70**, 1842-1844.

Pluske, J. B., Williams, I. H. and Aherne, F. X. (1996). Maintenance of villous height and crypt depth in piglets by providing continuous nutrition after weaning. *Animal Science*, **62**, 131-144.

Shankar, A. H. and Prasad, A. S. (1998). Zinc and immune function: the biological basis of altered resistance to infection. *American Journal of Clinical Nutrition*, **68**(supplement), 447S-463S.

Wilkie, B. and Mallard, B. (1999). Selection for high immune response: an alternative approach to animal health maintenance? *Veterinary Immunology and Immunopathology*, **72**, 231-235.

Williams, N. H., Stahly, T. S. and Zimmermann, D. R. (1997a). Effect of chronic immune system activation on the rate, efficiency, and composition of growth and lysine needs of pigs fed from 6 to 27 kg. *Journal of Animal Science*, **75**, 2463-2471.

Williams, N. H., Stahly, T. S. and Zimmermann, D. R. (1997b). Effect of level of chronic immune system activation on the growth and dietary lysine needs of pigs fed from 6 to 112 kg. *Journal of Animal Science*, **75**, 2481-2496.

Wong, K. F., Middleton, N., Montgomery, M., Dey, M. and Carr, R. I. (1998). Immunostimulation of murine spleen cells by materials associated with bovine milk protein fractions. *Journal of Dairy Science*, **81**, 1825-1832.

Yoo, S. S., Field, C. J. and McBurney, M. I. (1997). Glutamine supplementation maintains intramuscular glutamine concentrations and normalizes lymphocyte function in infected early weaned piglets. *Journal of Nutrition*, **127**, 2253-2259.

7   The Enemy Within: Non-Infectious Diseases and Oxidative Stress

As pointed out in the previous chapter the world is a dangerous place and animals are continually subject to attack by a multitude of pathogenic organisms. These may enter the body in feed, water or in the air and animals have of necessity evolved complex and successful systems to avoid these pathogens and to limit their effect after they may have invaded the body. In addition the medical and veterinary sciences of the last century have made great strides in developing drugs and vaccines that control many infectious organisms. It is clear also that many nutritional strategies can be developed that will mitigate the effect of the infectious disease pressure on animals raised for food.

Paradoxically however, the successful control of many infectious diseases has served to illuminate another problem, that of non-infectious, metabolic, or physiological diseases. These diseases are not caused by any infectious organism but are physiological responses to various environmental and nutritional stresses. The non-infectious diseases become more significant once the major infectious diseases are under control.

There are two major categories of non-infectious diseases, skeletal problems and diseases related to oxidative stress. Skeletal problems revolve around bone and joint development and are a major health issue for both ruminant and monogastric animals. Leg weaknesses and lameness afflict many animals raised under intensive conditions and this is an important welfare concern as well as having a serious impact upon animal growth and performance. Oxidative stress is associated with many non-infectious diseases and their control is based upon dietary intake of various antioxidant agents. This has been extensively discussed previously (Adams, 1999a).

Non-infectious diseases by their very nature cannot be controlled by antibiotics or other antimicrobial agents. These diseases are strongly influenced by diet. Consequently Total Nutrition must ensure that feed formulations and nutritional practice are designed preferably to avoid non-infectious diseases.

Bone and joint problems

A major source of non-infectious disease is related to problems of bone structure and skeletal development. It is now becoming increasingly obvious that rapid growth rates of animals raised for food and rapid deposition of muscle tissue frequently exceeds the capacity of the skeleton to support the animal. This is manifested in various bone and joint problems such as lameness in pigs and cattle, tibial dyschondroplasia and other leg weaknesses in broilers, tendon strain and damage in horses. A disease known as nutritional degenerative myopathy can occur in cattle on pasture in the spring. This is characterised by skeletal and cardiac problems leading to lameness and sometimes to sudden death (Walsh *et al.*, 1993) and antioxidant status of the animals seems to be important here.

Bone and joint problems also fuel the perception among the general population that the welfare status of the animals is being sacrificed for growth rate. However bone or joint problems do in fact have quite a serious economic impact upon animal production as well as impact upon the welfare status of the animals so it is very important to attempt to use nutrition to minimise these difficulties.

The skeleton is extremely important in protecting the internal organs such as the brain, spinal cord, heart and lungs. It is the attachment site for muscles and ligaments and supports movement of the body. The skeleton is also a metabolic reservoir for calcium and phosphorus. Poultry in particular have developed a specific type of bone, medullary bone, which can be quickly formed and resorbed in response to the extreme demand for calcium during egg laying. This medullary bone in avian species enables birds to store large amounts of calcium that is readily available to meet the demands of egg-shell formation. Bone is also an important source of calcium for foetal development and for milk production.

Bone is a living tissue which undergoes a continuous cycle of bone formation, involving osteocyte and osteoblast cells, and bone resorption involving oteoclasts. Bone is a very complex material consisting of collagen fibres, a mineral component, as crystals of calcium and phosphate, in the form of hydroxyapatite $(Ca_{10}(PO_4)_6(OH)_2$ and other ions and a ground substance formed by glycoproteins. There are two morphologically different forms of bone in the skeleton. These are cortical (compact) and cancellous or trabecular (spongy) bone. Cortical bone is dominant in the femur and tibia whereas spongy bone is dominant in the vertebrae and pelvis. Cortical bone provides rigidity and is mainly responsible for mechanical and protective functions. Spongy bone provides elasticity. It is metabolically more active than cortical bone and is responsible for approximately 50% of skeletal metabolism.

Normal bone growth and mineralization are the result of a complex interplay of genetic, cellular, hormonal and environmental influences. Disturbances at any level can result in bone abnormalities and disease. Mineralization of bone is important during growth as this gives good mechanical strength. Many animal breeding programmes are directed towards increase in growth rates of muscle and therefore bone mineralization rates must keep up if bone and joint problems are to be avoided.

Adequate bone development requires a certain amount of physical activity as well as good nutrition. It has long been known that vitamins C and D are very important for bone formation and animal diets must certainly be well supplied with these vitamins to avoid bone and joint problems. Another problem is osteoporosis which is characterised by a loss of bone from the skeleton and deterioration of bone structure. Osteoporosis is of primary importance in the ageing process, particularly in women and is probably not so very important in animals raised for food. Bone development is a complex biochemical process and seems to be associated with alkaline phosphatase activity, but the exact role of this enzyme is unclear. However alkaline phosphatase might have some future value as a means of assessment of Total Nutrition as further discussed in Chapter 8.

The skeleton is not stable in so much as large amounts of calcium and of phosphorus in bone can be liberated by resorption. Bone is continuously turned over in the body and this takes place particularly during lactation and egg production. Bone turn over or remodelling involves the replacement of a quantity of existing bone by new bone. During growth this process results in bone expansion with bone formation exceeding bone resorption. This bone turnover maintains the ability of the skeleton to withstand the physiological and mechanical demands placed upon it.

Calcium nutrition

Calcium is a curious essential element. It is the most abundant mineral element in the animal body. It is a major component of bone and egg shell and consequently is required in substantial amounts in all diets. However the absorption of calcium from the gastrointestinal tract is a major problem because many calcium salts are insoluble. A good example is fatty acids which form insoluble soaps with calcium in the gastrointestinal tract. This should not be a problem in healthy animals but may result in less calcium being available when lipids are malabsorbed. Oxalic acid and phytic acids, both of which occur in feed raw materials form insoluble complexes with calcium and this may inhibit absorption. Dietary fibre may also bind calcium although microbial fermentation in the large intestine may well release this calcium and it would become available again.

Various compounds such as lactose, gastric acids, bile acids and the active form of vitamin D_3 (1,25-dihydroxycholecalciferol) improve calcium absorption in the small intestine. Lactose, which is the major carbohydrate in milk, stimulates the absorption of calcium by interacting with the absorptive cells of the gastrointestinal tract to increase their permeability to calcium (Armbrecht and Wasserman, 1976). In the stomach some calcium is probably converted into the more soluble chloride form by reaction with hydrochloric acid. In the small intestine calcium may be kept in solution by reaction with bile acids. This may also be a beneficial effect of the use of organic acids in piglet diets. The presence of acid in the gastrointestinal tract should ensure that calcium is readily absorbed but this requires further investigation.

It seems that various non-digestible oligosaccharides have a general effect in stimulating calcium absorption (Van Loo et al., 1999). Non-digestible oligosaccharides (Chapter 5) are readily fermented by bacteria in the large intestine and this increases volatile fatty acid production. Improved calcium absorption in the large intestine may be related to this fatty acid production. The increased absorption occurs mainly in the large intestine and leads to improved bone mineral density. This introduces a new concept as it is generally accepted that calcium absorption occurs mainly in the small intestine.

Rats fed on a diet containing galacto-oligosaccharides absorbed calcium more efficiently than those on the control diet and had an improved calcium retention (Table 1) (Chonana and Watanuki, 1995). The bone ash weight and tibia calcium content of rats fed on galacto-oligosaccharides was significantly higher than those of control animals (Chonan et al., 1995). Other non-digestible oligosaccharides such as fructo-oligosaccharides also enhance calcium absorption (Morohashi et al., 1998), so the response is likely to be a general effect of non-digestible oligosaccharides.

Table 1
Effect of galacto-oligosaccharides (GOS) on the apparent calcium absorption and retention in rats

Calcium	Treatment		
uptake	Control	GOS (5%)	GOS (10%)
Absorption (%)	72.0a*	81.7b	86.3b
Retention (%)	71.9a	81.5b	86.1b

* Values having different letters in the same row differ significantly (P <0.05).

Various non-digestible oligosaccharides have been widely studied as prebiotics but now it is interesting to think of these materials perhaps as helping in the difficult task of calcium absorption. Feeding of various non-digestible oligosaccharides might have a useful benefit in improving bone calcification and consequently various skeletal problems of animals.

After absorption calcium is not freely mobile in the cytoplasm of cells (Trewavas, 1999). Calcium binds to many proteins that are attached to the cytoskeleton or to membrane surfaces. Other important intracellular calcium stores are the endoplasmic reticulum, the mitochondria and possibly the Golgi vesicles. There is some calcium remaining in the cytoplasm after protein binding and organelle uptake known as "resting" calcium. Calcium-dependent calcium-ATPases rapidly pump excess calcium into organelles and vesicles within cells to maintain a low calcium level in the cytoplasm. Calcium channels connect the stores of calcium in the organelles and vesicles with the cytoplasm and permit the flow of calcium between the cytoplasm and the other cellular components.

Calcium also plays many important roles in metabolism and is essential for the activity of many enzymes including phospholipase A_2 nucleases and a-amylase. It is involved in contractile properties of muscle and plays a role in blood clotting.

Calcium plays a central role in oxidative damage of mitochondria (Crawford et al., 1998). Oxidative stress leads to a release of calcium from intracellular endoplasmic reticular and mitochondrial stores. This calcium release results in an elevation of cytosolic levels which has been linked to cellular damage, activation of mitochondrial permeability transition protein (MPT), growth arrest and disease. Calcium may also be responsible for activating nucleases that directly degrade the mitochondrial RNA and DNA. Activation of calcium–dependent nucleases is a known consequence of cellular stress.

Calcium nutrition is of major importance for all species of animals. It also has evident welfare effects as well as serious implications for the economics of growth, lactation and egg production. Total Nutrition must ensure that calcium nutrition is fully adequate to meet the productive needs of the animal.

Assessment of calcium requirement is not easy and the concept of maximal calcium retention has been used as a functional indicator of calcium requirement in humans (Cashman and Flynn, 1999) and this could also be of relevance in animal nutrition. To achieve maximum skeletal strength it is important to develop and maintain a calcium reserve in the skeleton which depends upon an adequate calcium intake. Maximum calcium retention is achieved at a level of calcium nutrition where bone growth is maximized but this is not readily measured in farm animals. However at calcium intakes above the maximum requirement calcium may be absorbed but instead of being retained in the skeleton it is excreted in the urine and this could readily be measured.

Oxidative stress

Oxygen is clearly required for normal respiration in animals yet at the same time oxygen is potentially a toxic substance and this has frequently been described as the oxygen paradox (Miller and Brzezinska-Slebodzinska, 1993). During the normal respiration process of animals, oxygen is progressively reduced to yield water. However the incomplete reduction of oxygen during this process leads to the formation of chemical entities that have powerful oxidising properties. These are known as reactive oxygen species (ROS) and basal cellular metabolism in the body of an animal continuously produces ROS (Table 2). They are also produced by the immune system as part of the strategy to destroy invading micro-organisms. The production of ROS and the challenge they mount to the well-being of the animal is quite significant. As much as 1-4% of the oxygen consumed in metabolism is diverted into the generation of ROS (Boveris and Chance, 1973). Consequently living cells are continuously subjected to oxidative stress.

Table 2
Sources of reactive oxygen species in the body

Exercise
Ischaemia and reperfusion
Phagocytes
Peroxisomes
Metabolism of arachidonic acid
Mitochondria
Reactions with iron and other metals
Xanthine oxidase enzyme

These ROS may cause extensive oxidative damage to biological molecules. They cause peroxidation of polyunsaturated fatty acids in cellular membranes. They modify DNA in cells through base-alterations, single strand breaks of the DNA chain and generate DNA-protein cross links. Certain amino acid residues in proteins are also modified by ROS and this causes inactivation of important enzymes and makes the proteins more susceptible to destruction by proteolysis.

Oxidative stress occurs when there is an imbalance in the ration of ROS to antioxidant systems. The may be due to either an overproduction of ROS or a deficiency of the antioxidant systems. Various different stresses may cause an increase in ROS production. The stimulation of the immune system may lead to a massive local production of ROS. Molecules containing copper or iron will stimulate ROS production. A decreased level of the antioxidant vitamins C and E will lead to higher levels of ROS.

Production of ROS is stimulated by consumption of high energy diets. Diet restriction decreases oxidative stress and increases the life span in animals. However modern animal production inevitably requires the feeding of high density diets to achieve good levels of economic performance and this in turn will increase the risk of oxidative stress. Therefore a regular consumption of antioxidants in the diets is important to prevent or limit the deleterious effects of oxidative stress.

If there is a massive production of ROS it may cause cell death because of the irreversible degradation of important molecules such as DNA, and this can lead in turn to the death of the animal. If the production of ROS is more moderate it is frequently not lethal but generates an oxidative stress which can lead to the onset of various non-infectious diseases. The ROS may be detoxified by the various antioxidant systems.

Oxidative stress can repress the activity of T-cells and alter the immune response. It can slow down growth rate of cells and this may be of significant importance in animal production (Morel and Barouki, 1999). Oxidative stress may also be responsible for decreased body weight, muscle wasting and skeletal muscle abnormalities.

Oxidative stress due to the formation of ROS in the body, is already recognised as an important factor in the development of cardiac diseases and cancers, in the ageing process and in failures of the immune system (Matés *et al.*, 1999). There is increasing evidence that oxidative stress is related to a wide range of diseases syndromes in both humans and animals. There is serious concern that oxidative stress may trigger selective mutation of RNA viruses. In this scenario oxidative stress will be linked directly to some infectious diseases as well as non-infectious diseases.

Animals are constantly exposed to systems that generate ROS and damage proteins, nucleic acids and lipids. These include a number of environmental factors such as ultra-violet light, and atmospheric pollutants, ozone, smoke, and nitrogen dioxide. However many ROS are inevitable by-products of normal metabolic processes such as autoxidation of reduced forms of NADH, NADPH, reduced flavins, cytochrome P450, inflammatory reactions, nitric oxide synthesis, oxidase catalysed reactions, lipid peroxidation, glycation/glycoxidation reactions and metal catalysed reactions.

Elevated levels of ROS are found in body fluids and tissues in almost any diseases state in animals. However it is now realised that ROS do not necessarily cause disease but may be a consequence of the disease. This has important implications in health and nutrition. Supplementation of animal diets with a variety of antioxidants are unlikely to have a curative effect against existing diseases syndromes. However it seems clear that adequate levels of antioxidants in diets may prevent diseases syndromes from developing.

Oxidative stress produces many deleterious effects in cells and tissues including the oxidation of lipids, proteins, and DNA; depletion of cellular energy stores, release of intracellular calcium stores; and the disruption of important cellular processes.

Excessive generation of ROS compromises cellular defences against oxidative stress. This in turn can result in peroxidation of membrane lipids and accumulation of lipid peroxidation products such as malonyl dialdehyde (MDA), membrane dysfunctions, increased cellular permeability and reduced viability. Membrane dysfunction of endothelial cells (cells lining blood vessels and heart), either internal or at the cell surface, will result in profound changes in cellular structure, including inflammation, and a wide variety of hematological and cardiovascular complications.

The presence or absence of oxidative stress is related to the lipid levels and the antioxidant levels in the feed (Wang et al., 1996). Pigs fed on high levels of sunflower oil without vitamin E supplementation had a higher concentration of thiobarbituric acid (TBA)-reactive materials in the red blood cells than those animals with supplemental vitamin E (Table 3). Ethane and pentane levels in exhaled breath of the pigs without additional vitamin E was also considerably higher than in those animals fed with high levels of vitamin E. Hydrocarbons in exhaled breath and TBA-reactive substances are indicators of lipid peroxidation in tissues (Chapter 8). A vitamin E level of 20 mg/kg was adequate for a diet containing 3% lard but for diets containing 10% sunflower oil this level was insufficient. The requirement for vitamin E or possibly other antioxidants in young pigs increases as the polyunsaturated fatty acid (PUFA) levels in the diet increases.

Table 3
Concentration of TBA-reactive substances (μmol MDA / 100 g haemoglobin) from erythrocytes of pigs given diets with varying levels of supplementary vitamin E

Duration of Experiment (days)	Vitamin E supplementation (mg/kg feed)		
	0	20	100
35	15.81b*	8.39a	9.21a
42	12.42b	8.65a	9.23a
49	11.84b	9.21a	8.26a
56	14.23b	9.15a	9.25a

*Values having different letters in the same row differ significanty (P<0.05)

Reactive oxygen species (ROS) are also produced by various inflammatory cells, including macrophages, as a part of the immune defence system. The production and regulation of ROS from inflammatory cells is likely to have significant health consequences. Production of ROS as an integral part of the immune response is necessary to protect against invading pathogens and impaired production of ROS makes the animal more susceptible to infections. In contrast over production of ROS or reduced ROS scavenging by antioxidants is also undesirable and will result in damage to macromolecules and the onset of oxidative stress (Dietert and Golemboski, 1998). Clearly control of ROS production is vital for good health and protection against infectious diseases but at the same time they may be the causative agent of various non-infectious diseases. Healthy tissues are normally protected against oxidative stress by a host of antioxidant enzymes and by various cellular antioxidants. It is important that these antioxidant defence systems maintain normal cellular physiology by scavenging ROS, avoiding tissue damage and supporting the immune system.

Oxidative stress, embryo development and early growth in chicks

In the chick embryo and in newly hatched chicks there is considerable metabolism of unsaturated fatty acids which are prone to autoxidation. Consequently antioxidant systems are needed in the embryonic tissues and must maintain antioxidant protection during different stages of embryogenesis and in particular at hatching time when oxidative stress is highly likely.

The hatching and early neonatal stages are particularly critical periods in terms of the risk of oxidative stress. At hatching the chick is suddenly exposed to atmospheric oxygen and has a dramatic increase in metabolic rate. The brain of the day old chick is highly enriched in long chain polyunsaturated fatty acids (C_{20} and C_{22}), and the liver also has a large concentration of unsaturated fatty acids (Surai et al., 1996). Consequently there is a danger that the hatching process may subject the chick to severe oxidative stress which could cause irreversible damage to the brain and central nervous system. It is well known that vitamin E deficiency in the chicken induces encephalomalacia by peroxidative damage of the brain.

In newly hatched chicks the vitamin E content was dramatically depleted over the first nine days to about 5% of the initial level (Table 4), (Surai et al., 1998). The carotenoid content was also rapidly depleted from the liver during the initial nine days after hatching

Table 4
Concentration of vitamin E and of carotenoids (mg/g tissue) in liver of chicks

Chick age (days)	Vitamin E	Carotenoids
-1	489.3	48.2
+1	566.2	54.4*
+5	143.2*	37.2*
+9	26.7*	16.9*

* Denotes significant difference from preceding development stage.

It seems likely that carotenoids are an important part of the antioxidant system of the developing chick embryo which has a beneficial effect upon hatchability. There is considerable anecdotal evidence that poultry breeders fed maize-based diets yield better progeny than those on wheat-based diets. One major difference between a maize-based diet and a wheat-based diet is clearly the much higher carotenoid content in maize. There is also the question of which particular carotenoid would be important. In maize the major carotenoid is lutein with a small proportion of zeaxanthin. However the red carotenoids canthaxanthin and astaxanthin have good antioxidant properties and they may also be valuable.

Increased levels of carotenoids in the egg yolk have been associated with increased accumulation in the embryonic tissues protecting them from oxidative damage during oxidative stress (Surai and Speake, 1998). Hatching eggs and chick tissues obtained from hens fed high levels of carotenoids in the diet were significantly less susceptible to *in vitro* oxidation. This suggests that carotenoids in the hatching egg are able to mitigate oxidative stress on the developing chick.

Clearly in Total Nutrition both vitamin E and carotenoids will play an important role in breeding hen diets and in early chick nutrition in avoiding non-infectious diseases due to oxidative stress.

Pulmonary hypertension syndrome (ascites)

Ascites is a non-infectious disease in broilers and is a major cause of economic losses. The syndrome is a basic problem of oxygen supply and demand that develops in response to cardio-pulmonary insufficiency. The modern high performance broiler has a low proportion of lung volume to body weight and this is responsible for the inability of the respiratory system to respond to the demand for oxygen, which leads to an elevated blood pressure within the pulmonary circulation. This sets off a sequential development of hypoxia, right-sided congestive heart failure, cirrhosis of the liver, and leakage of fluids from blood vessels which accumulate in the body cavity and produce the ascites syndrome.

Ascites may be triggered by environmental factors such as low temperatures, lung damage or high altitude hypoxia. However it also occurs under optimal conditions simply in response to rapid growth. Frequently the respiratory system in unable to respond to the high oxygen needs of the rapidly growing broiler. This leads to hypoxia, respiratory acidosis, accumulation of fluids in the body cavity and ultimately death of the broiler. During hypoxia there is increase in the production of ROS leading to oxidative stress. Tissue damage caused by the ROS attracts white blood cells which in turn release more ROS causing further damage.

There has been considerable interest in the relationship between ascites and vitamin E in broilers. Lower levels of vitamin E were found in liver and lung tissues of broilers suffering from ascites compared to control birds (Enkvetchakul *et al.,* 1993). More detailed studies showed that the mitochondria from lungs of broilers suffering ascites were malfunctioning due to oxidative stress and this could be alleviated by high levels of dietary vitamin E (Iqbal *et al.,* 2001). Vitamin E has a beneficial effect in lowering oxidative stress in broilers with ascites. Unfortunately however feeding high levels of vitamin E is not always effective in alleviating the ascites problem (Bottje *et al.,* 1997).

Selenium together with high levels of vitamin E (250 IU:kg feed) also offers some useful protection against ascites (Roch *et al.,* 2000).

The ascites problem may actually be due to some metabolic defect in the mitochondria of susceptible birds (Cawthon, *et al.,* 2001). This defect would allow electrons to leak from the respiratory chain in the mitochondria and form an increased quantity of ROS and lead to oxidative stress. If this scenario is correct then good levels of dietary antioxidants may offer some protection against ascites but would by no means be a guaranteed cure.

There has also been some interest in the value of the amino acid arginine in alleviating ascites in broilers (Ruiz-Feria, *et al.,* 2001). Arginine is an essential amino acid for poultry because birds lack the enzyme carbamyl phosphate synthetase, which is involved in the conversion of ornithine to citrulline and onto arginine. Arginine is also an important amino acid in supporting the immune system (Chapter 6). However the efficacy of supplemental arginine in reducing the incidence of ascites in field trials has been highly variable.

Viruses

Oxidative stress in implicated in the pathogenesis of several viral diseases including hepatitis, influenza and AIDS (Beck and Levander, 1998). ROS are key participants in damage caused by viral infection such as inflammation of epithelial cells. Dietary oxidative stress caused by either selenium or vitamin E deficiency transforms a normally benign RNA virus, coxsackievirus B3, to a virulent form which causes heart damage in both humans and mice (Beck, 1999). The mechanism seems to be that the nucleotide sequence in the benign virus changes under oxidative stress to resemble the genome of the virulent strains. The fact that both selenium and vitamin E deficiency in mice resulted in similar problems of viral infection strongly suggests that this is a general non-specific phenomenon of oxidative stress and not something specifically related either to the properties of selenium or of vitamin E. It has been widely recognised for many years that poor nutrition increases susceptibility to disease but now there is the rather dramatic possibility that poor nutrition can actually generate a disease. The possibility that other viral diseases may emerge in response to poor nutrition must seriously be considered.

Feed intake and oxidative stress

Intake of high density diets is related to oxidative stress and feed restriction of animals reduces oxidative stress (Sohal and Weindruch, 1996). This is not a practical possibility for food animal production where high density diets are necessary to promote rapid growth and support economic performance. Nevertheless it will always be important to ensure that high density diets are well protected with antioxidants and that feeds are not manufactured from oxidised materials.

Oxidative stress is likely to be related to the consumption of oxidised feeds as well as to events within the body of the animal. The gastrointestinal tract will be the first organ in the body to suffer damage from oxidised feed ingredients and this could have an impact upon nutrient absorption and resistance to pathogenic micro-organisms through effects upon the immune system. Feeding oxidised fats to broilers increased turnover of enterocyte cells in the gastrointestinal tract and of liver cells and gave an overall reduced performance (Dibner *et al.,* 1996). An increased turnover of enterocytes is also associated with reduced nutrient absorption capacity and this would inevitably result in poorer growth performance. An increased proliferation of liver cells in both pigs and broilers fed oxidised fats was not influenced by the addition of an antioxidant to the oxidised fat. This is a good example of the virtue of prevent of oxidation versus attempts to cure the problem afterwards. Presumably liver cell proliferation is caused by cytotoxic compounds generated during oxidation of the fat and these are not affected by the presence of antioxidants in the feed. It is extremely important to ensure that the feed is not oxidised by use of appropriate antioxidants.

Immune protection against pathogenic micro-organisms that live in the gastrointestinal tract or that may arrive in feed is accomplished in part by secretion of the immunoglobulin IgA into the tissues and lumen of the gastrointestinal tract. Feeding oxidised fats to broilers may lead to reduced disease resistance through a reduction in secretion of IgA or through reduced stability of IgA. In pigs, there is a reduction in proliferation of lymphocytes which produce the IgA in response to the feeding of oxidised fats. The overall effect is a reduction in the efficiency of the immune system associated with the gastrointestinal tract. Consequently oxidative stress could then be manifested as increased susceptibilities to infectious diseases.

Antioxidant systems

Animals have developed a battery of antioxidant agents and systems that converts ROS to unreactive entities and thus avoid cellular damage by these processes. Biological antioxidants act at three levels (Table 5).

These are: (1) naturally-occurring enzymes and proteins which can prevent the uncontrolled formation of ROS, (2) molecules that inhibit the reaction of ROS with other cellular components and (3) systems that repair damaged molecules (Chaudière and Ferrari-Iliou, 1999).

Table 5
Antioxidant defence systems in living cells

Defence mechanism	Molecular basis
Inhibition of formation of ROS	Superoxide dismutase Glutathione peroxidase Catalase Metal-binding proteins
Restriction of propagation and chain reactions	Carotenoids Glutathione Polyphenols Uric acid Vitamins A, E and C
Repair of damaged molecules	Lipases Proteases

The molecule, glutathione (GSH) acts as an antioxidant buffer in cells as it is ubiquitous. Adequate concentrations of glutathione are required to for a variety of immune functions including lymphocyte activation, natural killer cell activation, and lymphocyte-mediated cytotoxicity (Chapter 6). GSH exists in animal cells almost entirely as an intracellular component. The primary function of GSH in the cell is to act as a regulator of the oxidation-reduction potential of the cell as an antioxidant. GSH is specifically involved in maintaining sulphhydryl enzymes in their active reduced states.

Several enzymatic systems are also able to detoxify ROS. Superoxide dismutase (SOD) eliminates the superoxide species and generates hydrogen peroxide which is broken down by catalase. Glutathione peroxidase reduces peroxides. Quinone reductase and haem oxygenase can prevent the formation of ROS.

Various metal-binding proteins such as ceruloplasmin, ferritin and transferrin have an important antioxidant function. Many important metabolites such as, $NADP^+/NADPH$, $NAD^+/NADH$, lipoic acid, uric acid and bilirubin act as intercellular antioxidants. There are also many components of the diet such as vitamins A, C and E, carotenoids, polyphenols and zinc which are important antioxidants. A number of antioxidant enzymes also require trace elements such as selenium, copper, zinc and iron from the diet as cofactors. Dietary nutricines will contribute to the overall redox balance as a source of natural antioxidants and the amount of oxidation that occurs will reflect the balance between the prooxidant and antioxidant activities that will be dictated by the particular environmental and dietary factors. The antioxidant nutricines must now be considered major factors in health maintenance and disease avoidance

Antioxidants from plants

Many molecules in plants are known to have an antioxidant activity. This includes various herbs and spices such as tumeric and rosemary, carotenoid pigments and a host of polyphenolic compounds. There is increasing interest in animal nutrition in the use of various plant extracts. However, frequently the antioxidant effect has only been demonstrated *in vitro*. This indicates that materials could have an application for the preservation of feeds from oxidative spoilage but it does not indicate that they would exert any effect upon the animal in terms of controlling oxidative stress.

The bioavailability and persistence of feed-borne antioxidants will be limited by their sensitivity to destruction in the gastrointestinal tract and to their ability to be absorbed into the bloodstream. It is possible to study the *in vivo* antioxidant effects of various plant extracts by measuring peroxidation of phospholipids in membranes of red blood cells and of lipids in the liver of mice (Asai *et al.*, 1999). Feeding mice extracts of tumeric, rosemary and capsicum reduced the susceptibility of red blood cell membranes and of liver lipids to oxidation. It is possible to influence the oxidative status of animals by including various antioxidants in the diet.

Oleuropein, an antioxidant molecule from olive oil, reduces the susceptibility of low density lipoproteins (LDL) to oxidation in rabbits (Coni *et al.*, 2000). In this work rabbits were either given feed containing olive oil (10%) or containing oleuropein (7mg/kg). In both treatments there was a reduced susceptibility to oxidation of the low density lipoproteins confirming that supplementation of feed with natural antioxidants can have beneficial effects in the body of the animal.

Polyphenolic nutricines

Plants produce thousands of phenolic compounds and consequently they are ubiquitous in feed materials of plant origin. Considerable attention is now being focussed on the flavonoids which includes a large number of compounds; chalcones, flavones, flavanones, flavanols and anthocyanins (Duthie *et al.*, 2000). These polyphenolic nutricines are bioactive substances widely occurring in feed and therefore potentially available for absorption into the blood. They are effective antioxidants in a wide range of oxidation systems scavenging peroxyl radicals, superoxide, hydroxyl radicals, nitric oxide and peroxynitrites. These are all important ROS implicated in many diseases syndromes, and so the ability of dietary nutricines to control these ROS will probably be important in health maintenance.

Malignant transformation is routinely demonstrated in cultured animal cells as an increase in uncontrolled cell growth. Many plant polyphenols are at least equal to, or in many cases superior to other dietary protection agents such as ß-carotene or vitamin E in mitigating this transformation. The citrus flavonoids hesperetin and hesperidine were among the most potent, inhibiting malignant transformation almost completely at very low concentrations (1μM) (Duthie *et al.*, 2000).

Flavones and flavanones in the diet increase the hepatic glutathione transferase (GST) activity in rats (Siess *et al.*, 1996). GST induction is generally considered to reflect an increase in cellular protection, ensuring that potential toxins are conjugated and excreted more rapidly.

Caffeic acid is an abundant polyphenolic material. It is one of the hydroxy cinnamic acids and is found naturally in many plants. It has powerful antioxidant activity as an oxygen radical scavenger and a chain-breaking antioxidant. It also acts synergistically with α-tocopherol, both delaying α-tocopherol consumption and recycling α-tocopherol from the α-tocopheroxyl radical during lipoprotein oxidation. Caffeic acid can be taken up by live cells without any cytotoxic effect (Nardini *et al.*, 1998). Caffeic acid treated cells had an increased resistance to oxidative stress. This was probably due to the ability of caffeic acid to reduce the loss of glutathione and to inhibit lipid peroxidation.

Altogether these results indicate that caffeic acid exerts an antioxidant action inside the cell and is able to modulate the response to an oxidative challenge, likely through the modulation of the cellular redox balance. These data support the view that dietary phenolic antioxidants other than vitamin E may play a role in the modulation of oxidative processes *in vivo*.

Carotenoids

Carotenoids are the most numerous and widespread group of pigments in nature and have long been used as colouring agents in the poultry, aquaculture and food manufacturing industries (Adams, 1999b). Carotenoids have a range of different colours from a light golden yellow of lutein to a dark red of canthaxanthin. Animals in general cannot synthesise the carotenoid molecules but obtain significant quantities in the diet from various plant sources. They play very important roles in the photosynthesis process in plants but their physiological role in animal tissues is still not very clear.

Some carotenoids are precursors of vitamin A but probably fewer than 10% of the carotenoids can be metabolised to vitamin A. Only ß-carotene, α-carotene and ß-cryptoxanthin are major vitamin A precursors in animals. However in modern animal nutrition vitamin A is added into the feed with other vitamins and carotenoids would only supply a small proportion of the vitamin A requirement. Recently however the antioxidant properties of carotenoids have been recognised.

There is also evidence that carotenoids have wide ranging effects in controlling several non-infectious disease syndromes (Basu et al., 2001). Good levels of carotenoid intake in feeds is possibly quite important in the promotion of immunocompetence. High dietary intake of carotenoids enhances the ability of T-lymphocytes to proliferate and this improves the immune response.

Dietary lutein obtained from the marigold flower reduced mammary tumour growth and development in mice (Park et al., 1998). Tumour incidence, weight and volume were generally lower in lutein fed mice. This suggests that lutein was capable of both preventing tumour initiation and of inhibiting tumour growth.

Lutein has been strongly implicated in protection of the eye against age-related macular degeneration and cataracts (Johnson, 2000). Lutein may act in the eye as a blue-light filter, protecting the underlying tissues from phototoxic damage.

The antioxidant activity of carotenoids may be quite important in hatching eggs and in the viability of newly hatched chicks as discussed above. It seems quite possible that inclusion of various carotenoids in the diet of food animals could be beneficial in addition to the desirable pigmenting properties they show.

Trace metals

The trace metals selenium, copper, iron and zinc are important in oxidative stress. Copper and iron are powerful prooxidants and therefore their amount in tissues has to be strictly controlled. Selenium and zinc have recognised antioxidant functions. Consequently in Total Nutrition attention must also be focused on ensuring a correct mineral balance in the feeds.

Copper may also be implicated in the so-called "prion" diseases or transmissible spongiform encephalopathies (TSEs). The TSEs include scrapie in sheep, bovine spongiform encephalopathy (BSE) in cattle and Creutzfeldt-Jakob disease (CJD) in humans (Cooper, 2001). The normal prion protein (PrP) in the brain binds copper and this copper binding is associated with superoxide dismutase activity, which is an important antioxidant enzyme. This antioxidative effect may be the normal role of the prion protein. The altered form of the prion protein (PrP^Sc) which is associated with the development of the diseases does not seem to bind copper.

Manganese can also bind to normal prion protein but this reduces the superoxide dismutase activity. Manganese may be able to convert the normal prion protein into the abnormal form and the prevalence of manganese in the soil may be related to the spontaneous form of the TSE. Studies with mice have shown that in the absence of the normal prion protein there is an increased oxidation of lipid and protein in all tissues (bio-Klamt *et al.*, 2001). It is possible that at least a part of the problem of prion diseases results from a disruption of a copper-dependent antioxidant enzyme system. During the development of prion diseases the whole organism becomes more sensitive to ROS injury, leading to a progressive disruption of tissues and vital organs, especially the central nervous system.

Zinc is now considered as an antioxidant in addition to its vital roles as a co-factor in many enzymes. Zinc has been used in industrial processes for many decades to galvanise iron and steel and prevent their oxidation. In this process the iron or steel is coated with a thin layer of zinc and this very significantly retards oxidation by the air, otherwise known as rusting.

Zinc in the form of zinc-bishistidinate has a distinct cardio-protective effect against ischaemic injury (Powell, 2000). Zinc is quite valuable in human applications in attenuating ischaemic and post-schaemic injury of cardiac tissue. Zinc is also capable of inhibiting the process of protein oxidation. A protein that has been oxidatively modified is subject to rapid destruction by cellular proteolytic enzymes and zinc may play an important role in protecting proteins from oxidation and subsequently from proteolytic destruction. This suggests that adequate levels of zinc nutrition may be vital in avoiding oxidative stress.

Future directions

Organic acids, non-digestible oligosaccharides and calcium nutrition

There is still a need to improve uptake of calcium to support the rapid growth and high levels of performance required for low cost food production. Various organic acids either added directly to feed or produced by fermentation of non-digestible oligosaccharides in the large intestine possibly aid in calcium absorption. Fumaric acid for example, at 2% in the diet of weaner pigs improved the balance of several minerals including calcium (Kirchgessner and Roth, 1980). A consensus report on the functional food properties of non-digestible oligosaccharides concluded that there was promising evidence that consumption of inulin-type fructans resulted in increased calcium absorption (Van Loo *et al.*, 1999).

There is clearly scope for much further research on the interactions between various organic acids, non-digestible oligosaccharides and fermentation activity in the gastrointestinal tract in relation to calcium nutrition.

Oxidative stress

Oxidative stress and the role of antioxidants in non-infectious diseases has become a major research area in both human and animal medicine and nutrition. However there is also considerable evidence that oxidative stress exerts a very fundamental role in metabolism by repressing gene expression (Morel and Barouki, 1999). This gives an even greater role for antioxidants in helping to restore the correct regulation of gene expression. Indeed this implies that there is the possibility of a nutritional control of gene expression.

Oxidative stress and mycotoxins

Ochratoxin A is produced by certain *Penicillum* and *Aspergillus* species of storage fungi and may contaminate feeds when it causes serious losses in productivity of animals. One mode of action of the toxicity of ochratoxin A is the generation of lipid peroxides in both rats and chicks (Hoehler *et al.*, 1997). Further work with rats indicated that ochratoxin A caused a 22% decrease in α-tocopherol blood plasma levels and a five-fold increase in the oxidative stress protein haem oxygenase-1 specifically in the kidney (Gautier *et al.*, 2001). These results strongly suggest that ochratoxin A generates an oxidative stress which may contribute to the kidney damage associated with the toxicity of ochratoxin A.

Aflatoxin production by the toxigenic fungus, *Aspergillus paraciticus*, seems related to oxidative stress and indeed may be a prerequisite for aflatoxin synthesis. A natural antioxidant, eugenol inhibits aflatoxin production (Jayashree and Subramanyam, 2000).

These observations offer the interesting possibility that it could be possible to control aflatoxin production and mycotoxin toxicity through antioxidants. This could be very important in feed and raw material conservation programmes. It would also be useful to determine if other mycotoxins were also synthesised under conditions of oxidative stress.

Oxidative stress and viruses

It has been widely recognised for many years that poor nutrition renders animals more susceptible to various diseases. However now there is evidence that the nutrition of the host may actually generate a pathogenic form of a virus (Beck and Levander, 1998).

There is great potential danger here as RNA viruses in particular have high mutation rates, high yields and short replication times (Domingo and Holland, 1997). RNA viruses are quite important pathogens of animals and include the causal agent of foot and mouth disease and swine influenza. Mutation rates are certainly affected by the environment. About 50 new viruses have emerged over the last 20 years, and most of them are RNA viruses. Consequently we are likely to be faced with a continuous evolution and emergence of new virus pathogens. The relationship between these new pathogens and nutrition will be of enormous importance if we are to successfully combat these new emerging diseases.

This also raises the question as to how many outbreaks of nutritional deficiency diseases may actually be the result of a virus that has changed it's pathogenic characteristics in response to replicating in a nutritionally deficient host. There is an urgent need to try and understand more clearly the relationship between viral diseases and the nutritional status of the host. This has powerful implications for the concept of Total Nutrition where we are concerned with health maintenance and disease avoidance. For example, will high levels of dietary antioxidants prevent the transformation of viruses into pathogenic strains ? Total Nutrition may be able to play a valuable role in modulating the emergence of pathogenic viruses once we understand more completely the relationship between nutritional status and viral pathogenesis. There is clearly much exciting nutritional research to be done in this area.

Trace metals and bovine spongiform encephalopathy (BSE)

If BSE and the human equivalent CJD is related to manganese and copper interactions (Cooper, 2001) then mineral nutrition will become even more important in controlling both health and growth of animals. BSE may be related to oxidative stress (bio Klamt et al., 2001) and this perhaps is induced or exacerbated by dietary manganese intake. It will be important to establish these links more firmly and to establish to what extent high levels of dietary antioxidants could protect animals from prion diseases. The concept of oxidative stress and antioxidant nutricines could well become of major importance in animal nutrition.

Feed and antioxidants

Feed composition is a major factor under our control which can affect the antioxidant status of animals. Fat content, both quantity and type may change the amount of fatty acids susceptible to oxidation. Shifting from more saturated animal fats to polyunsaturated plant oils will alter the oxidative status of the animal and may well have an impact upon flavour and shelf-life of meat products. Mineral nutrition will also be important.

Polyphenolic nutricines

These have always been a significant part of the diet of animals but more information is needed on the bioavailability and absorption of these compounds into the bloodstream. Gut micro-organisms might destroy or modify the polyphenols before they can be absorbed. However, polyphenols have been detected in the plasma and urine of rats and in humans various flavonoids are absorbed from food (Hollman and Katan, 1999), so it is highly likely that they could play a useful role in controlling oxidative stress. Also any evidence of toxicity of polyphenols would need to be determined.

Assessment of oxidative stress

Oxidative stress is implicated in a whole host of disease syndromes, which may include BSE. Consequently it will be important to be able to measure the degree of oxidative stress in an animal and to relate this to nutrition. This however is not a trivial task as the various chemical species involved in oxidative stress such as the different free radicals and oxidised macromolecules are not easy to analyse. Furthermore for any assessment system to be of practical value it must be relatively cheap, simple and not require sacrifice of the animal. There are many possibilities being developed. Breath may be analysed for volatile hydrocarbons, ethane and pentane. Measurement of thiobarbituric acid reacting substances in blood is another possibility. Isoprostanes are molecules formed from oxidation of arachidonic acid in cellular membranes and are excreted in urine. The amount of isoprostanes in urine is a reflection of oxidative stress of the host animal. The assessment of oxidative stress and its relationship to health and nutrition is an important topic and is more fully discussed in Chapter 8.

The production of ROS by inflammatory cells is beneficial as part of the immune response but excessive production will lead to oxidative stress which is undesirable. This raises the question as to what levels of ROS production and subsequent scavenging by dietary antioxidants and antioxidant enzymes are necessary to avoid disease and maintain optimum health in the animal. Probably different animal species will require a different balance of dietary antioxidants to maintain health and productivity. There is little information available on this topic and it will not be easy to investigate. However it will be important in the future to understand the relationship between various dietary antioxidants and health. In obese strain chickens prone to the development of autoimmune thyroiditis, dietary antioxidants delayed the onset of the disease (Bagchi *et al.*, 1990). This is an interesting example of the ability of dietary antioxidants to avoid disease and maintain health which is the central thesis of Total Nutrition and much further research needs to be undertaken to establish the relationship between nutrition, disease avoidance and health maintenance.

References

Adams, C. A. (1999a). *Nutricines. Food components in health and nutrition*. Nottingham University Press, UK. pp. 18-33.

Adams, C. A. (1999b). *Nutricines. Food components in health and nutrition*. Nottingham University Press, UK. pp. 114-116.

Armbrecht, H. J. and Wasserman, R. H. (1976). Enhancement of Ca^{++} uptake by lactose in the small intestine. *Journal of Nutrition*, **106**, 1265-1271.

Asai, A., Nakagawa, K. and Miyazawa, T. (1999). Antioxidative effects of turmeric, rosemary and capsicum extracts on membrane phospholipid peroxidation and liver lipid metabolism in mice. *Bioscience Biotechnology and Biochemistry*, **63**, 2118-2122.

Bagchi, N., Brown, T. R., Hergeden, T. M., Dhar, A. and Sundich, R. S. (1990). Antioxidants delay the onset of thyroiditis in obese-strain chickens. *Endocrinology*, **127**, 1590-1595.

Basu, H. N., Del Vecchio, A. J., Flider, F. and Orthoefer, F. T. (2001). Nutritional and potential disease prevention properties of carotenoids. *Journal of the American Oil Chemists Society*, **78**, 665-675.

Beck, M. A. (1999). Selenium and host defence towards viruses. *Proceedings of the Nutrition Society*, **58**, 707 – 711.

Beck, M. A. and Levander, O. A. (1998). Dietary oxidative stress and the potentiation of viral infection. *Annual Review of Nutrition*, **18**, 93-116.

Bio Klamt, F., Dal-Pizzol, F., Conte da Frota Jr., M. L., Walz, R., Andrades, M. E., da Silva, E. G., Brentani, R. R., Izquierdo, I. and Moreira, J. C. F. (2001). Imbalance of antioxidant defense in mice lacking cellular prion protein. *Free Radical Biology and Medicine*, **30**, 1137-1144.

Bottje, W. G., Erf, G. F., Bersi, T. K., Wang, S., Barnes, D. and Beers, K. W. (1997). Effect of dietary dl-α-tocopherol on tissue α- and gamma-tocopherol and pulmonary hypertension syndrome (ascites) in broilers. *Poultry Science*, **76**, 1505-1512.

Boveris, A. and Chance, B. (1973). The mitochondrial generation of hydrogen peroxide. *Biochemical Journal*, **143**, 707-711.

Cashman, K. D. and Flynn, A. (1999). Optimal nutrition: calcium, magnesium and phosphorus. *Proceedings of the Nutrition Society*, **58**, 477-487.

Cawthon, D., Beeers, K and Bottje, W. G. (2001). Electron transport chain defect and inefficient respiration may underlie pulmonary hypertension syndrome (ascites)–associated mitochondrial dysfunction in broilers. *Poultry Science*, **80**, 474-484.

Chaudière, J. and Ferrari-Iliou, R. (1999). Intracellular antioxidants: from chemical to biochemical mechanisms. *Food and Chemical Toxicology*, **37**, 949-962.

Chonana, O. and Watanuki, M. (1995).Effect of galactooligosaccharides on calcium absorption in rats. *Journal of Nutritional Science and Vitaminology*, **41**, 95-104.

Chonan, O., Matsumoto, K. and Watanuki, M. (1995). Effect of galactooligosaccharides on calcium absorption and preventing bone loss in ovariectomized rats. *Bioscience, Biotechnology and Biochemistry*, **59**, 236-239.

Coni, E., Di Benedetto, R., Di Pasquale, M., Masella, R., Modesti, D., Mattei, R. and Carlini, E. A. (2000). Protective effect of oleuropein, an olive oil biophenol, on low density lipoprotein oxidizability in rabbits. *Lipids*, **35**, 45-54.

Cooper, C. (2001). Metals and disease. Copper on the brain. *The Biochemist*, **23**, 11-13.

Crawford, D. R., Abramova, N. E. and Davies, K. J. A. (1998). Oxidative stress causes a general, calcium dependent degradation of mitochondrial polynucleotides. *Free Radical Biology and Medicine*, **25**, 1106-1111.

Dibner, J. J., Atwell, C. A., Kitchell, M. L., Shermer, W. D. and Ivey, F. J. (1996). Feeding of oxidised fats to broilers and swine: effects on enterocyte turnover, hepatocyte proliferation and the gut associated lymphoid tissue. *Animal Feed Science Technology*, **62**, 1-13.

Dietert, R. R. and Golemboski, K. A. (1998). Avian macrophage metabolism. *Poultry Science*, **77**, 990-997.

Domingo, E. and Holland, J. J. (1997). RNA virus mutations and fitness for survival. *Annual Review of Microbiology*, **15**, 151-178.

Duthie, G. G., Duthie, S. J. and Kyle, J. A. M. (2000). Plant polyphenols in cancer and heart disease: implications as nutritional antioxidants. *Nutrition Research Reviews*, **13**, 79-106.

Enkvetchakul, B., Bottje, W., Anthony, N., Moore, R. and Huff, W. (1993). Compromised antioxidant status associated with ascites in broilers. *Poultry Science*, **72**, 2272-2280.

Gautier, J-C., Holzhaeuser, D., Markovic, J., Gremaud, E., Benoit, S. and Turesky, R. J. (2001). Oxidative damage and stress response from ochratoxin a exposure in rats. *Free Radical Biology and Medicine*, **30**, 1089-1098.

Hoehler, D., Marquardt, R. R. and Frohlich, A. A. (1997). Lipid peroxidation as one mode of action in ochratoxin A toxicity in rats and chicks. *Canadian Journal of Animal Science*, **77**, 287-292.

Hollman, P. C. H. and Katan, M. B. (1999). Dietary flavonoids: intake, health effects and bioavailability. *Food and Chemical Toxicology*, **37**, 937-942.

Iqbal, M., Cawthon, D., Wideman, Jr., R. F. and Bottje, W. G. (2001). Lung mitochondrial dysfunction in pulmonary hypertension syndrome. II. Oxidative stress and inability to improve function with repeated additions of adenosine diphosphate. *Poultry Science*, **80**, 656-665.

Jayashree, T. and Subramanyam, C. (2000). Oxidative stress as a prerequisite for aflatoxin production by *Aspergillus paraciticus*. *Free Radical Biology and Medicine*, **29**, 981-985.

Johnson, E. J. (2000). The roe of lutein in disease prevention. *Nutrition and Clinical Care*, **3**, 289-296.

Kirchgessner, M. and Roth, F. X. (1980). Digestibility and balance of protein, energy and some minerals in piglets given supplements of fumaric acid. *Zeitschrift für Tierphysiologie, Tierernährung und Futtermittelkunde*, **44**, 239-246.

Matés, J. M., Pérez-Gómez, C. and De Castro, I. N. (1999). Antioxidant enzymes and human diseases. *Clinical Biochemistry*, **32**, 595-603.

Miller J. K. and Brzezinska-Slebodzinska, E. (1993). Oxidative stress, antioxidants, and animal function. *Journal of Dairy Science*, **76**, 2812-2823.

Morel, Y. and Barouki, R. (1999). Repression of gene expression by oxidative stress. *Biochemical Journal*, **342**, 481-496.

Morohashi, T., Sano, T., Ohta, A. and Yamada, S. (1998). True calcium absorption in the intestine is enhanced by fructooligosaccharide feeding in rats. *Journal of Nutrition*, **128**, 1815-1818.

Nardini, M., Pisu, P., Gentili, V., Natella, F., Di Felice, M., Piccolella, E. and Scaccini, C. (1998). Effect of caffeic acid on *tert*-butyl hydroperoxide-induced oxidative stress in U937. *Free Radical Biology and Medicine*, **25**, 1098-1105.

Park, J. S., Chew, B. P. and Wong, T. S. (1998). Dietary lutein from marigold extract inhibits mammary tumour development in BALB/c mice. *Journal of Nutrition*, **128**, 1650-1656.

Powell, S. R. (2000). The antioxidant properties of zinc. *Journal of Nutrition*, **130**, 1447S-1454S.

Roch, G., Boulianne, M. and de Roth, L. (2000). Dietary antioxidants reduce ascites in broilers. *World Poultry*, **16**, 18-22.

Ruiz-Feria, C. A., Kidd, M. T. and Wideman, Jr., R. F. (2001). Plasma levels of arginine, ornithine, and urea and growth performance of broilers fed supplemental L-arginine during cool temperature exposure. *Poultry Science*, **80**, 358-369.

Siess, M. H., Mas, J. P., Canivenc-Lavier, M. C. and Suschetet, M. (2000). Time course of induction of rat hepatic drug-metabolizing enzyme activities following dietary administration of flavonoids. *Journal of Toxicology and Environmental Health*, **49**, 481-496.

179

Sohal, R. S. and Weindruch, R. (1996). Oxidative stress, caloric restriction, and aging. *Science*, **273**, 59-63.

Surai, P. F., Noble, R. C. and Speake, B. K. (1996). Tissue-specific differences in antioxidant distribution and susceptibility to lipid peroxidation during development of the chick embryo. *Biochimica et Biophysica Acta*, **1304**, 1-10.

Surai, P. F., Ionov, I. A., Kuchmistova, E. F., Noble, R. C. and Speake, B. K. (1998). The relationship between the levels of α-tocopherol and carotenoids in the maternal feed, yolk and neonatal tissues: comparison between the chicken, turkey, duck and goose. *Journal of the Science of Food and Agriculture*, **76**, 593-598.

Surai, P. F. and Speake, B. K. (1998). Distribution of carotenoids from the yolk to the tissues of the chick embryo. *Journal of Nutritional Biochemistry*, **9**, 645-651.

Trewavas, A. (1999). Le calcium, c'est la vie: calcium makes waves. *Plant Physiology*, **120**, 1-6.

Van Loo, J., Cummings, J., Delzenne, N., Englyst, H., Franck, A., Hopkins, M., Kok, N., Macfarlane, G., Newton, D., Quigley, M., Roberfroid, M., van Vliet, T. and van den Heuvel, E. (1999). Functional food properties of non-digestible oligosaccharides: a consensus report from the ENDO project (DGX11 AIR11-CT94-1095). *British Journal of Nutrition*, **81**, 121-132.

Walsh, D. M., Kennedy, S., Blanchflower, W. J., Goodall, E. A. and Kennedy, D. G. (1993). Vitamin E and selenium deficiencies increase indices of lipid peroxidation in muscle tissue of ruminant calves. *International Journal of Vitamin and Nutritional Research*, **63**, 188-194.

Wang, Y. H., Leibholz, J., Bryden, W. L. and Fraser, D. R. (1996). Lipid peroxidation status as an index to evaluate the influence of dietary fats on vitamin E requirements of young pigs. *British Journal of Nutrition*, **75**, 81-95.

8 Monitoring Performance: Assessment of Total Nutrition and Feeding Standards

Total Nutrition has to fulfil several requirements. Feed must be produced that is safe and free from contamination by pathogenic micro-organisms. This feed must provide adequate levels of nutrients to prevent any overt deficiency diseases and to support rapid growth and high levels of productivity of animals. This has been one of the major achievements of modern animal nutrition and diets now can be routinely formulated to avoid deficiency problems. The nutrients must be readily consumed, easily digested, and rapidly absorbed to avoid enteric disorders. Feed must deliver adequate amounts and types of nutrients and nutricines to develop and support the immune system to control the incidence of infectious diseases. Feed must also contain adequate levels of nutricines to maintain good health and avoid non-infectious diseases and create minimum environmental damage. These requirements are of concern both for human nutrition and for the nutrition of animals raised for food.

There may well be different threshold levels of nutrients and nutricines required to fulfil these various functions. A threshold level of a nutrient or a nutricine will lie between the minimum requirement and the maximum tolerance. The minimum requirement will be the smallest amount that an animal must consume to avoid deficiency symptoms. This is only applicable to essential nutrients. The maximum threshold level will be the largest amount an animal can consume without an adverse effect. All dietary components, both nutrients and nutricines, will have a maximum threshold level at which they will lead to a nutritional imbalance or even toxicity. It will now be necessary to define the amount of nutrients and nutricines to support growth and productivity of animals, to maintain health and to avoid disease. This is the basic objective of Total Nutrition.

The conventional concept of levels of essential nutrients is based on observational and experimental findings that nutrients function to prevent deficiency diseases. In Total Nutrition many feed ingredients such as the nutricines will not have a minimum essential level based on avoidance of deficiency symptoms, as do common nutrients.

In Total Nutrition the concept of essential feed components recognises feed as being much more than a collection of nutrients; it is also an important source of nutricines which impact upon the health of our commercial animals.

The assessment of Total Nutrition will be a combination of several factors:

(1) nutrients levels required to avoid deficiency and support growth
(2) the amounts of nutrients or nutricines required to have an effect upon biomarkers or functional indices
(3) the levels of nutrients and nutricines which prevents disease.

In the new concept of Total Nutrition, the minimal level of any feed component, nutrient or nutricine that affects the metabolism and gastrointestinal function in a manner beneficial to good health and to animal performance, must be considered. It is clear that some feed components, nutricines, will be required in substantial amounts yet not have a direct nutrient function. For example organic acid nutricines are widely incorporated into piglet feeds at up to 1.5% to support good piglet health and performance. However no minimum level of these acids can be specified based on avoidance of deficiency symptoms as they are not classic essential nutrients. Similarly antioxidant nutricines that protect feed quality and avoid oxidative stress will not have a minimum level designated by the conventional concept of avoidance of deficiency symptoms. Both organic acid and antioxidant nutricines however are an important part of Total Nutrition and this has major implications for the concept of levels of essential nutrients and diet formulation.

Total Nutrition is not concerned solely with avoidance of nutrient deficiency symptoms, although this clearly must also be included as one component of Total Nutrition. The future strategy is no longer to define optimum nutrient intakes but rather to determine the Total Nutrition necessary to give optimum health and nutrient status in the animal. If the amount of nutrients and nutricines cannot be assessed through the establishment of levels to avoid deficiency, then an alternative concept will need to be developed.

Total Nutrition could be assessed by measurement of various functional indices of the target animal which should be directly related to disease mechanisms or to ill health (Strain, 1999). The use of functional indices requires a knowledge and understanding of the functions of nutrients and nutricines at the physiological level. A functional index for Total Nutrition will be a biochemical or physiological factor that can be measured and that is related to some function in the target animal. It should be influenced by changes in dietary intake or body stores of the nutrient or nutricine of interest. Total Nutrition would be achieved when the particular functional index is no longer affected by intake of the nutrient or nutricine concerned. This would cover both classic deficiency symptoms and any toxic symptoms.

The functional indices need to be of a general nature that will reflect disease symptoms or health status. Examples of general functional indices useful for indicating optimum nutritional status and thus setting the limits for Total Nutrition are given in Table 1.

Table 1
Possible functional indices for Total Nutrition

Oxidative stress and antioxidant status
Immune function
Calcium and bone health
Muscle quality
Nutritional stress proteins
DNA damage and repair
Urinary nitrite and nitrate
Volatile fatty acids in gastrointestinal tract

Oxidative stress and antioxidant status

Oxidative stress as discussed in Chapter 7 plays an important role in the development of many non-infectious diseases. It strongly influences the immune system which in turn will affect the ability of the animal to grow rapidly and to resist infectious diseases. Oxidative stress also influences the pathogenicity of viruses where benign forms of a virus may mutate into virulent forms in an animal under oxidative stress (Beck, 1999).

Antioxidant status is the physiological balance between the various antioxidant systems and the rate of production of pro-oxidants in the living animal. In mammals this balance is probably weighted in favour of oxidation as this is essential for the release of energy from feed. The metabolic utilization of nutrients to supply energy is in reality a series of carefully controlled oxidation reactions. A part of the immune response is also the generation of reactive oxygen species (ROS) and these have the ability to damage both pathogens and host tissues. Consequently the body has evolved various internal antioxidant systems to redress this oxidative balance and to avoid excessive oxidative stress.

The antioxidant status will also be influenced by dietary supply of antioxidants and feed quality directly affects the antioxidant status of an animal in both a positive and negative manner. Many components of feeds are known to have potent antioxidant activities and feed formulations with high levels of these materials will strengthen the antioxidant status of the animal. By contrast some feed ingredients such as polyunsaturated fatty acids and metals such as copper and iron may be easily oxidised or act as pro-oxidants. Disease conditions, environmental conditions such as heat stress or medication that depress feed intake will also affect the supply of dietary antioxidants. High quality palatable feed must be an initial requirement in Total Nutrition.

Evaluation of oxidative stress and antioxidant status is an important aspect in relating nutrition to health. It could be used to follow the health progress of animals and to adjust feed formulations when appropriate. There are two basic principle strategies to evaluate oxidative stress and antioxidant status. The antioxidant capacity of the animal body can be measured through the TEAC system or by determining various antioxidant enzymes or compounds such as glutathione or carotenoids. Alternatively the end products of oxidative stress can be measured. These can be aldehydes in the TBARS system, hydrocarbons, isoprostanes or uric acid. Antioxidant status is usually assessed by measuring some parameters in blood, urine or breath. There are several possible methods to assess oxidative stress in living animals as listed in Table 2.

Perhaps the oxidative status of the contents of the gastrointestinal tract should also be considered. From the time of ingestion of feed to the excretion of faecal material feed components undergo major physical and chemical changes in the gastrointestinal tract. In addition the microflora in the gastrointestinal tract may produce free radicals or antioxidants. The multitude of complex reactions occurring in the gastrointestinal tract could lead to the production of free radicals and oxidised end products which may have an effect upon the development of enteric disorders. The absorption of oxidised end products such as aldehydes and ketones may subsequently cause oxidative stress. The oxidative stability of the contents of the gastrointestinal tract may be an important factor in Total Nutrition. Perhaps the oxidative status of faecal material in animals might be a useful indicator of oxidative stress.

Table 2
Procedures to
assess oxidative
stress

Trolox equivalent antioxidant capacity (TEAC)
Ferric reducing ability of plasma (FRAP assay)
Antioxidant enzymes; glutathione peroxidase, superoxide dismutase, catalase
Antioxidants; carotenoids, glutathione, selenium, tocopherols, uric acid, vitamin C, zinc,
Thiobarbituric–acid reactive substances (TBARS)
Hydrocarbons in breath; ethane, pentane
LDL oxidation
Isoprostanes in blood or urine

Trolox equivalent antioxidant capacity (TEAC)

The TEAC assay is a spectrophotometric technique based on scavenging of long-lived radical anions that are generated through peroxidase activity of metmyoglobin in the presence of hydrogen peroxide (Rice-Evans et al., 1995). The scavenging activity of compounds in biological fluids such as blood plasma is compared to that of Trolox, a water-soluble vitamin E derivative. Under standardized conditions mixtures of antioxidants show an additive effect and so the assay can measure a mixture of antioxidants which is likely to be found in blood plasma for example (van den Berg, 1999).

Ferric reducing ability of plasma (FRAP assay)

This is a relatively simple automated colorimetric test, which measures the ability of blood plasma to reduce ferric to ferrous ions (Benzie and Strain, 1996). Results using this assay on blood plasma from Chinese males showed the FRAP values of healthy individuals to be around 1 mmol/litre. Similar studies do not seem to have been done on animals but this system seems particularly suitable to assess the oxidative status of animals as well as humans.

Antioxidant enzymes

Superoxide dismutase (SOD) catalyzes the dismutation of the reactive superoxide radical to produce hydrogen peroxide as shown in reaction (1). These enzymes are widely distributed in all living organisms and play an important role in removing the very reactive superoxide radical.

$$2O_2^{\cdot -} + 2H^+ \longrightarrow H_2O_2 + O_2 \qquad (1)$$

Hydrogen peroxide is then removed by two other enzymes, glutathione peroxidase (GSHPx) and catalase. Glutathione peroxidase converts hydrogen peroxide to water using glutathione as a hydrogen donor as illustrated in reaction (2). Glutathione (GSH) is a tripeptide composed of cysteine, glutamic acid and glycine and it is readily converted into an oxidized form (GSSG) where two molecules of glutathione are joined together forming oxidised glutathione (GSSG). The oxidised glutathione is reduced back to the GSH form by another enzyme, glutathione reductase.

$$H_2O_2 + 2GSH \longrightarrow GSSG + 2H_2O \qquad (2)$$

Glutathione peroxidase is a very important antioxidant enzyme because it can react with hydroperoxides from fatty acids or from cholesterol to form stable hydroxy lipids that do not decompose to form radicals or aldehydes that could cause cellular damage.

Catalase is another enzyme able to remove hydrogen peroxide (reaction 3) and is particularly high in the liver and red blood cells of animals.

$$2 H_2O_2 \longrightarrow 2H_2O + O_2 \qquad (3)$$

Whilst these antioxidant enzymes are very important in living cells they are not easily used as markers or indicators of oxidative stress.

Antioxidants

Carotenoids

Carotenoids are important nutricines in avoidance of non-infectious diseases as discussed in Chapter 7. They are also valuable in feeds for poultry breeders. Carotenoids such as lutein and zeaxanthin, which occur in many feed ingredients, can be detected in the blood of the target animals. This has already been utilised in some instances to follow the degree of pigmentation of broilers while they are growing. Yellow broilers must have a high level of carotenoids in the feed and this must then be absorbed from the gastrointestinal tract and transferred to the subcutaneous fat layers. To successfully pigment broilers they must have adequate levels of carotenoids in the blood and this is readily determined.

Hatching egg quality in breeders is also most likely related to the amount of carotenoids in the egg yolk. It is possible to observe higher carotenoids in the blood plasma of chick embryos derived from hens fed a high carotenoid diet than a control diet (Surai and Speake, 1998). In particular for the high-carotenoid group the concentration of carotenoids in the plasma of the embryos increased dramatically between days 19 and 22 of development.

Glutathione

Glutathione is metabolically very important as discussed above and is found in tissues of all animals. Because it is so important the amounts in tissues are highly regulated and therefore it is difficult to measure depletion or excess of glutathione in the body. Consequently glutathione levels are unlikely to be a useful way to assess oxidative stress or antioxidant levels in living tissues. However in normal living cells the ratio of the GSH/GSSG is greater than 10 so it might be possible to use this ratio as an indication of oxidative stress.

It functions as antioxidant as described above and is involved in the regeneration of oxidised vitamin E and in the removal of hydrogen peroxide. Glutathione is also intimately involved in supporting the immune system as discussed in Chapter 6.

Selenium

Selenium is a very important constituent of the active site of the antioxidant enzymes glutathione peroxidase which catalyzes the oxidation of hydroperoxides. Selenium and vitamin E may also act together with one sparing any deficiency of the other. Selenium status of animals can be measured in blood. Selenium has also been implicated in defence against viruses (Beck, 1999). Selenium is a recognised nutrient and adequate levels in feeds often depend upon the selenium status of the soils where the feed was produced.

Tocopherols

Tocopherols occur as different homologs (α, ß, γ, and δ) and these are readily assayed by HPLC techniques. The α-tocopherols have greater vitamin E activity than antioxidant activity whereas the γ- and δ-tocopherols have a more powerful antioxidant effect. They may also act synergistically with ascorbic acid.

Low levels of tocopherols in lung and liver tissue of broilers were associated with incidences of ascites (Enkvetchakul *et al.*, 1993).

Uric acid

Uric acid is produced by the oxidation of xanthine and hypoxanthine by the enzyme s xanthine oxidase and dehydrogenase and is excreted by avian species including poultry. In mammals the uric acid is also present in the blood but is further transformed into urea which is excreted.

Uric acid has an antioxidant function and in blood it may stabilise ascorbic acid by binding pro-oxidant metals such as copper and iron. It can also react with a number of reactive oxygen species such as peroxynitrite and the hydroxyl radical (OH). High levels of uric acid in the blood of humans can crystallize and cause the disease of gout. Consequently there has been little interest in raising levels of uric acid in the blood of humans. However it might be beneficial for animals where the life span is much shorter and it could be interesting to investigate further the relationship between uric acid levels and oxidative stress.

Vitamin C

Vitamin C or ascorbic acid is both a water-soluble vitamin and an important intercellular antioxidant and consequently it is difficult to separate the two functions. It regenerates vitamin E from its chromanoxyl radical or oxidised form.

Zinc

Zinc is an important micro-nutrient with several physiological functions. It plays an important role in the immune system and is also considered as an intercellular antioxidant. Zinc plays an indirect role as an antioxidant since it does not interact directly with free radicals. Rather it alleviates oxidative stress (Chapter 7), but it is nevertheless a very important nutrient and the zinc status of an animal could be an indication of good health and nutrition.

A major problem is the availability of a reliable functional index of zinc status. A whole range of indicators of zinc status have been suggested (Salguerio *et al.*, 2000). These include blood serum or plasma levels of zinc, zinc levels in cells of the immune system such as leukocytes or nutrophils, or the concentration of the metallothionein proteins in red blood cells. The metallothioneins are a group of small metal-binding proteins which are induced in response to zinc and have antioxidant effects (Powel, 2000).

Thiobarituric acid reactive substances (TBARS)

The TBARS assay has been widely used to measure oxidative stability in feeds and foods as well as a marker for oxidative stress. The assay is a colorimetric procedure where thiobarbituric acid reacts with malondialdehyde. However malodialdehyde may also be generated during inflammatory reactions in blood and the TBARS assay may not be completely representative of oxidative stress. Also the reaction is not specific and a large number of secondary oxidation products can generate the colour. Nevertheless it has the attraction of being relatively simple and inexpensive to carry out and has given useful information.

Ascites syndrome is an important economic disease in poultry and seems to be associated with oxidative stress. The liver and hearts of broilers with ascites showed an increase in the concentration of TBARS indicating a high level of lipid oxidation in these organs (Table 3), (Diaz-Cruz *et al.*, 1996). It seems that values of around 0.4 nmol/mg protein were normal and in broilers with ascites this went up to around 1.0 nmmol/mg protein. It would be interesting to establish if these values were also observed in other cases of oxidative stress.

Table 3 Levels of thiobarbituric acid reactive substances (TBARS) in liver and heart of broilers with and without ascites

Tissue	Ascites	TBARS (nmol/mg protein)
Liver	No	0.40
	Yes	0.95
Heart	No	0.42
	Yes	1.12

Hydrocarbons in breath

Oxidation of polyunsaturated fatty acids in the body produces many diverse end products which include the hydrocarbons ethane and pentane and a portion of these is released into the breath. Samples of breath can be collected and analysed by gas chromatography (GC). Increased levels of hydrocarbons in blood of rats can be measured after the animals have been subjected to an oxidative stress (Dillard *et al.*, 1977). Both the type and amount of polyunsaturated fatty acids (PUFA) in the feed of laboratory animals influence the production of hydrocarbons in breath (Kneepens *et al.*, 1994). It is also influenced by dietary intakes of antioxidants such as vitamin E, selenium, ß-carotene and vitamin C. It is a relatively simple technique and is worthy of further consideration in animal nutrition.

LDL oxidation

An increased level of low-density lipoprotein (LDL) cholesterol is considered a major risk factor for development of atherosclerosis in humans. Oxidised LDL results in foam cell formation which in turn leads to atherosclerosis. Consequently in human nutrition there is considerable interest in measuring LDL oxidation status. However this does not have much relevance for animal production and study of LDL oxidation is not a trivial matter.

Isoprostanes

Isoprostanes are prostaglandin-like compounds that are produced in mammals from free radical-catalysed oxidation of fatty acids such as arachidonic acid. Furthermore isoprostanes are dramatically induced by oxidative stress (Lawson *et al.*, 1999). Isoprostanes are chemically quite stable end products of lipid oxidation and they are very useful as a marker of oxidative stress *in vivo*. Isoprostanes are considered to be the most reliable and most accurate markers of oxidative stress. Oxidative stress increases in cases of asthma in humans and this was also associated with an increase in isoprostanes in blood plasma (Wood *et al.*, 2000). Isoprostane levels increase in experimental models of injury and can be suppressed using antioxidants (Morrow and Roberts, 1997; Roberts and Morrow, 1997).

Unfortunately assay of isoprostanes is not a trivial task so its widespread use as a marker of oxidative stress must await the development of easier and cheaper assay systems. An immunoassay method (ELISA test) and a GC/MS (gas chromatography/mass spectrometry) method have been developed but both are relatively expensive and labour intensive procedures.

The routine measurement of antioxidant nutricines and oxidative status may become more common in the future with the increasing development of automated analytical systems. Assays of selenium and isoprostanes are always likely to be expensive and labour-intensive. However there are already routine methods available for carotenoids, vitamin E, ascorbic acid, and the TBARS, TEAC and FRAP assays. Antioxidant enzymes such as superoxide dismutase, glutathione peroxidase and catalase are also readily assayed.

No single simple method is available at the present time to provide an accurate estimation of antioxidant status. Consequently several different methods must be selected, each of which has it own attractions and limitations. The precise definition of optimal antioxidant status is not yet established although it is generally recognised that feeds should contain adequate antioxidant nutricines to avoid oxidative stress. Some antioxidants such as vitamin E, selenium and vitamin C are essential nutrients and these have recommended minimum values for general nutrition. However whether this level will be suitable to guard against oxidative stress is not yet established. Most of the dietary antioxidants either synthetic such as BHT or BHA or natural products such as carotenoids and flavonoids are not essential nutrients and have no officially recommended minimum levels.

Some nutricines such as carotenoids and antioxidants have maximum permitted levels at which they may be incorporated in feeds. For example under current EU legislation carotenoids may be added to poultry feeds up to a maximum of 80 mg/kg and synthetic antioxidants such as BHA or BHT may be used at levels of 150 mg/kg of feed. These levels have certainly not been set on the basis of any assessment of oxidative stress but rather on the basis of toxicological safety.

Immune status

A strong and responsive immune system is important to maintain good animal health. Both the nutritional status of an animal and specific nutrients and nutricines may affect the immune system directly by activating immune cells or altering immune cell interactions. It is now widely recognised that there is an interaction between adequate nutrition and the functioning of the immune system.

However assessment of immune function is not an easy task and at present there is no overall measure of immunity in animals. This makes it difficult to determine the effect of a nutricine or nutrient upon the immune system. Furthermore nutritional status is unlikely to influence only one part of the immune system but will probably influence several parts of the immune system. Consequently it will be difficult to assess studies aimed the development of immunomodulators and immunostimulants

The objective in considering immune status in Total Nutrition is to determine whether a particular nutrient, nutricine, or feed formulation will improve immune function and influence immunity to infectious diseases. The ideal scenario would be to use various indices of immunology to predict resistance to infection and to detect instances of poor immune function. Two basic methods are available, *in vivo* and *in vitro* procedures.

In vivo assessment of immune status

A useful *in vivo* procedure is antibody response to vaccines. Immunization with appropriate antigens will elicit the production of serum antibodies and therefore an assay for specific antibodies could provide information about resistance of animals to infectious diseases.

In vitro assessment of immune status

In vitro procedures are based on isolation of immune cells from the animal body. These cells are then subjected to a variety of tests to measure their proliferative ability or the release of molecules such as cytokines. Proliferation of immune cells is commonly determined by measuring the uptake of compounds such as thymidine that are radioactively labelled. Such assays however require highly trained personnel and are by no means simple to carry out.

An alternative *in vitro* system is to study the uptake of glucose and glutamine which are the major energy sources for cells of the immune system. When immune cells are activated, use of these nutrients increases substantially and immune cells are extremely sensitive to changes in availability of energy sources (Wu *et al.*, 1991).

An interesting possibility is to measure the amount of nitrite which is derived from nitric oxide produced from arginine by the enzyme nitric oxide synthase. Immune cells such as macrophages cultured from spleen of chickens challenged with *Eimeria acervulina* produced increasing amounts of nitrite as the coccidiosis developed. A similar effect was seen in chickens suffering from poult enteritis and mortality syndrome (PEMS) where again nitrite production by macrophages increased (Qureshi *et al.*, 1998).

Antisecretory factor

Newborn piglets have little immunity towards enteric diseases and a positive transfer of protection from the sow's milk is crucial to their survival. Antisecretory factor is a protein which has been found in sow's milk and is able to protect the piglet against diarrhoea (Lönnroth *et al.*, 1988). A daily intake in sow's milk of 1mg was sufficient to prevent diarrhoea in piglets. Consequently it may be necessary in the future to monitor sow performance to ensure that suitable protective factors are transferred in the milk.

Calcium and bone health

Calcium is a major component of bone tissue and most of the body calcium is located in bone. However calcium levels are also important in blood, muscles and other tissues where it is important in mediating vascular contraction and vasodilation of the blood supply network and in nerve transmission. Other key nutrients such as potassium, magnesium, fibre, ß-carotene and vitamin C also play an important role in maintaining bone health (New *et al.*, 2000).

Good skeletal strength depends upon an adequate calcium reserve in the body and calcium retention rate has been suggested as a useful parameter in humans (Jackman *et al.*, 1997). This attempts to assess the lowest value of calcium intake at which maximum calcium retention is obtained.

Measurement of bone mass could be another important functional indicator of Total Nutrition. This requires special equipment and is carried out in studies on humans (New *et al.*, 2000); however it is unlikely to be widely used in general animal nutrition work.

Biochemical functional indices of bone turnover may also provide useful information on the health status of animals. Turnover of bone tissue is essential to maintain good skeletal strength. Bone turnover is an active process whereby bone tissue is resorbed and new tissue synthesised. It is the balance between bone formation and bone resorption that determines the bone mass and thereby skeletal strength. This is extremely important in all classes of livestock where leg weaknesses and lameness can cause serious loss in animal performance.

Biochemical indices of bone turnover rely on the measurement of enzymes in the urine or blood. Hydroxyproline is an amino acid found in urine and has been related to bone resorption. Pyridinoline and deoxypyridinoline in urine have been used as markers of bone resorption. Total alkaline phosphatase activity and osteocalcin in blood serum are widely used in clinical practice as indices of bone formation (Robins and New, 1997, New *et al.*, 2000).

Muscle quality

Muscle quality is an important physiological characteristic of animals. Many animals are raised for meat and muscle quantity and quality is of obvious importance. Other animals such as high producing dairy cows often have difficulty in eating sufficient feed to support milk production and body proteins may be catabolized to support milk production. It would clearly be valuable to be able to assess the protein status of animals in relation to nutrition and there are some interesting possibilities here.

Meat quality

One of the main objectives in raising animals for food use is to provide a source of meat, which in physiological terms is muscle tissue. Good muscle quality is obviously important in maintaining health and productivity of animals but also is extremely important in determining final meat quality. Over the years commercial breeding programmes for meat animals have largely focused on improvements in growth rate and meat yield. Poultry have also been selected for increased production of breast muscle as a proportion of the total carcass. These improvements in productivity will have an effect upon muscle structure and function.

Meat quality is determined by complex interactions of several different factors. These include nutrition and general husbandry, stresses on the animals immediately prior to slaughter and handling of the carcass after slaughter. There are two recognised extremes in meat quality: PSE (pale, soft exudative) and DFD (dark, firm and dry). The meat quality is influenced by the size of the muscle tissue since large muscle mass chills more slowly, which results in low pH due to glycolysis, and producing PSE-type meat. Large muscle fibres, which occur in breast tissue of modern turkeys for example, also have more glycolytic activity than smaller muscle fibres. This also can lead to lower pH and PSE-type meat.

Muscle damage may occur in the body due to injury or perhaps mycotoxins in the feed and this can be assessed by measurement of the enzyme creatine kinase in the blood. This enzyme may leak into the blood stream if there is an oxidative stress or injury or weakness in the cell membranes of the muscle tissue and so is an indicator of good muscle integrity (Lauridsen *et al.*, 1996). In broilers the activity of plasma creatine kinase decreased with increasing levels of vitamin C in the diet. This probably indicates an improved membrane stability in the muscle tissue. However studies with pigs fed combinations of vitamin E and copper showed no differences in the creatine kinase levels in the blood (Lauridsen *et al.*, 1999). Creatine kinase is potentially a useful parameter as it can be easily determined using commercially available test kits and so is worthy of further study in relation to health and nutrition.

Many of the problems of meat quality are associated with nutrition and slaughter practices. Nutritional supplements may help alleviate the symptoms of PSE (pale soft exudative) meat. Antioxidants will protect against muscle damage and help improve shelf-life of the meat.

Muscle breakdown

In dairy cows nutrient intake shortly after calving is often insufficient to meet the increased energy and protein requirements resulting from the initiation of milk production. Consequently cows mobilize body proteins, which includes the catabolism of muscle proteins. The muscle proteins actin and myosin contain the unusual amino acid 3-methylhistidine. This is not re-utilized in protein synthesis but is excreted in the urine where it can be readily measured. Therefore urinary excretion of methylhistidine can be used to assess muscle protein breakdown in cattle. High yielding dairy cows showed a significant increase in urinary excretion of methylhistidine indicative of increased catabolism of muscle proteins after calving (Table 4). This is mainly due to a difference between nutrient intakes and nutrient requirements during early lactation (Plazier *et al.*, 2000). This is another possible marker to assess the nutritional status of animals.

Table 4
Live weight and excretion of 3-methyl histidine in pre-and postcalving dairy cows

Dairy cow	Live weight (kg)	3-methyl histidine (mmol/day)
Pre-calving	737	2.48a*
Post-calving	697	4.11b

*values with different letters significantly different (P<0.001)

Nutritional stress proteins

Nutritional stress on an animal leads to the synthesis of various proteins which can be detected in the tissues of the body and could act as a biomarker of the nutritional status of the animal. Some proteins which may be of interest in Total Nutrition are acute phase proteins and Bcl-2.

Acute phase proteins

These are a diverse group of proteins synthesised in the liver during the early phase of a stress response (Figure 1). This stress can arise from inflammation of tissues, infections, onset of disease or environmental stress such as transport. Synthesis of acute phase proteins is a protective response by the animal and these proteins appear in the blood earlier than specific antibodies.

There are several different acute phase proteins that have been identified in cattle, pigs, and poultry (Table 5).

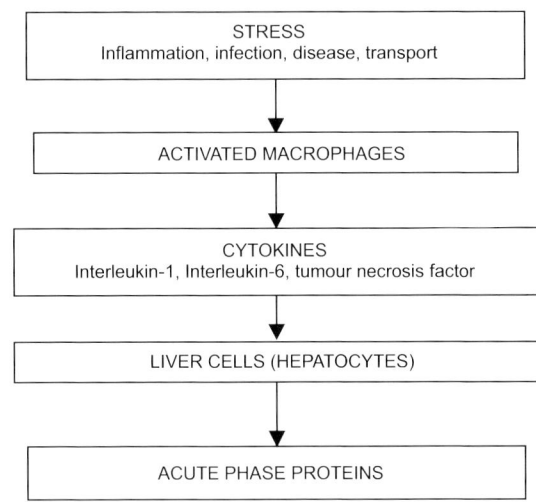

Figure 1
Chain of events for the synthesis of acute phase proteins

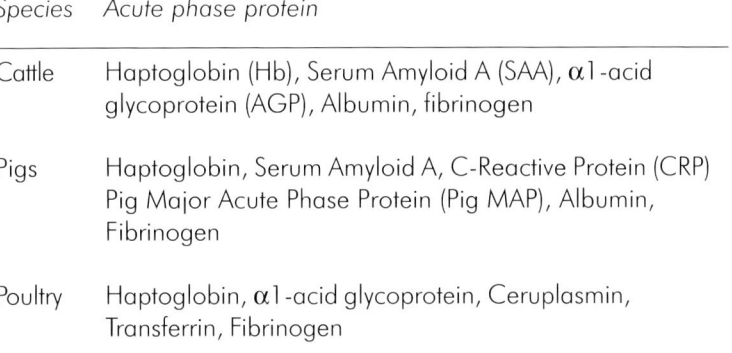

Table 5
Acute phase proteins identified in various animal species

Species	Acute phase protein
Cattle	Haptoglobin (Hb), Serum Amyloid A (SAA), α1-acid glycoprotein (AGP), Albumin, fibrinogen
Pigs	Haptoglobin, Serum Amyloid A, C-Reactive Protein (CRP) Pig Major Acute Phase Protein (Pig MAP), Albumin, Fibrinogen
Poultry	Haptoglobin, α1-acid glycoprotein, Ceruplasmin, Transferrin, Fibrinogen

The acute phase protein response is also associated with alterations in plasma mineral levels including the withdrawal of zinc from the blood plasma into the liver and the release of copper from tissue stores into the blood. These events are linked to growth depression and decreased production (Chamanza et al., 1999).

Acute phase proteins are plasma proteins which increase in concentration in response to various stresses. Consequently assay of these proteins in blood can provide an objective measure of the health status of the animal. They are increasingly being used as markers of animal health and welfare and would be a very useful way to assess Total Nutrition.

At the present the measurement of acute phase proteins in blood plasma is still largely limited to research laboratories. However various test kits are now available which makes it feasible to assay acute phase proteins in a routine manner to assess nutrition and health status of animals under practical conditions.

Bcl-2 protein

Cultured cells of the human colon produced elevated levels of Bcl-2 protein *in vitro* in response to a nutritional stress (Singh and Paraskeva, 1998). This may be associated with the development of colorectal cancers as cells of the colon switch their energy metabolism from using butyric acid to using glucose in tumour development. However the synthesis of such a protein might be a useful indicator of a more general nutritional stress.

DNA damage and repair

Damage to deoxyribonucleic acids (DNA) is involved in at least two major human problems, ageing and cancer. These are not very important in the raising of animals for food, although problems of ageing are of interest in pets.

Various reactive oxygen species (ROS) and free radicals produced in cells of the animal body continually damage DNA and this must be repaired. Damage to DNA is manifested by chemical changes in the four bases, adenine, cytosine, guanine and thymine which make up the DNA the DNA molecule. The most common change is the conversion of guanine to 8-hydroxy guanosine and this may be used as an index of oxidative damage to DNA.

Several of the oxidised molecules from DNA damage are excreted in the urine in humans in the form of nucleosides, the bases linked to the sugar deoxyribose. The nucleoside 8-hydroxy-deoxyguanosine can be measured using HPLC techniques. In a study with humans smokers excreted 50% more 8-hydroxy-deoxyguanosine than non-smokers suggesting a 50% increase in the rate of oxidative damage to their DNA from smoking (Loft *et al.*, 1993).

Another system called "Single Cell Microgelelectrophoresise" or "COMET" assay has also been developed which involves studying the DNA in cells of animals and this can be done with blood samples (Fairbairn *et al.*, 1995). This is a more rapid technique than analysis of altered DNA bases and also reveals DNA damage due to toxic materials as well as oxidative damage. This could well have application in animal nutrition and health.

Further work in this area needs to be done on animals to see if oxidative stress can be detected in this way. Again the procedures to measure compounds such as 8-hydroxyguanosie are fairly complex and more rapid and simpler methods will be needed for widespread application in animal nutrition.

Urinary nitrite and nitrate

Urinary nitrite and nitrate is derived from nitric oxide (NO), which is produced in the body by cells such as macrophages when activated by infecting bacteria. The nitric oxide is rapidly oxidised to nitrite and nitrate and excreted in the urine. Measurement of urinary nitrite + nitrate levels has been used as a quantitative biomarker to assess total intestinal bacterial infections (Bovee-Oudenhoven et al., 1997).

The basic principle is that urine samples are firstly stabilised against bacterial deterioration by addition of an antibiotic to the urine samples. The nitrate in the samples is usually then chemically reduced to nitrite and the total nitrite determined by a colorimetric procedure. Automated analytical systems are available for this type of approach and so it is feasible for more widespread use in assessing the status of animal health and nutrition.

In rats orally infected with Salmonella the levels of nitrite + nitrate increased in the urine over a six day period until it was some five-fold higher than in the control rats (Bovee-Oudenhoven et al., 1999). This procedure clearly has interesting applications in food animal production as infection by Salmonella is of major concern in terms of safe food and it might be possible to track the infection status of animals by measuring the nitrite + nitrate levels in urine samples.

Volatile fatty acids in the gastrointestinal tract

Short chain volatile fatty acids (SCFAs), mainly acetic, propionic and butyric are produced by microbial fermentation in the gastrointestinal tract. Analysis of the digesta from the small and large intestines can give a profile of SCFAs and this is influenced by diet and health status of the animal.

In general there should be low levels of SCFAs in the small intestine as these are an indication of microbial fermentation and excessive microbial colonisation of the small intestine is not desired but may occur under conditions of nutritional stress (Bedford, 2000). Fermentation of resistant starch and NSP normally occurs in the large intestine and this can be a useful way to recover energy from undigested carbohydrate feed components.

Measurement of SCFAs is a relatively simple technique although using digesta from the small intestine would require the sacrifice of animals. However analysis of SCFAs in faecal material would be much simpler and may also be related to animal health and performance.

Characteristics of feed formulations in total nutrition

Feeds must be produced which are nutritionally adequate and economically feasible. This requires attention to the storage and stability of raw materials and the finished feeds. Feeds can be assessed in terms of antioxidant and microbiological status and in amounts of non-starch polysaccharides. These are all important characteristics of feeds used in Total Nutrition where feed quality becomes even more crucial in supporting health and avoiding disease.

Total antioxidant status of feed

Antioxidants play two important roles in animal health and nutrition. Firstly they protect the nutritional quality of feeds against autoxidation and secondly they impact upon animal health and performance through influencing the antioxidant status of the animal as discussed in Chapter 7.

One possibility of assessing this is to assess feeds by the spectrophotometric technique known as the Trolox equivalent antioxidant activity (TEAC) (Rice–Evans et al., 1995). In this assay the antioxidant capacity of feeds are compared to that of Trolox which is a water-soluble analogue of vitamin E. This assay has been widely used in human nutrition (Rice-Evans and Miller, 1996), and could be adapted to animal feeds.

Microbiological status of feeds

In most countries feed must be free from *Salmonella*, *Clostridia* and other pathogenic bacteria. Total *Enterobacteriacea* may well be used as a criterion of microbiological quality in the future although there is debate about what levels should be set as a minimum. Values from 10-1000 CFU/g have been suggested. In practice 10 CFU/g is probably too low for a routine standard and 100 CFU/g is more realistic.

Mould contamination is a particular danger in stored raw materials. Moulds are ubiquitous contaminants of natural materials such as cereal grains and it will be difficult to eliminate all mould contamination. However the use of organic acid nutricines as discussed in Chapter 2 will make a marked improvement in feed quality. There are no universally agreed methods to determine mould contamination and consequently no universally agreed levels of moulds which may be permitted. However it is quite clear that to achieve Total Nutrition mould must be reduced as far as practicable. The secondary metabolites of moulds, the mycotoxins, are also of major concern in animal health and performance. Most countries have stringent control on levels of aflatoxin permitted in raw materials and animal feeds. In Europe however aflatoxin is not much of a problem whereas Ochratoxin A, zearalenone and deoxynivalenol (DON) are probably of more consequence. In Germany maximum levels of 1mg/kg of DON and 0.05 mg/kg of zearalenone have been set as specifications for various cereals.

Non-starch polysaccharides (NSPs)

Non-starch polysaccharides encompass a wide range of different molecules and have a wide range of effects including changes in gut physiology, fat metabolism glucose uptake and enteric disorders as discussed in Chapters 4 and 5. The NSPs are usually classified into soluble and insoluble fractions. The soluble forms will exert their effect in the small intestine increasing viscosity of the digesta and reducing fat digestion. The insoluble forms or the fibre fraction will exert their effect in the large intestine and influence fermentation by the resident microflora and also faecal bulk and transit time.

Analysis of NSPs in feeds can be done but is neither rapid nor simple and consequently is unlikely to come into practical use as a routine feed quality parameter.

Nutrient utilization

Modern animal nutrition has developed the concept of high quality dense diets. These are usually formulated to minimum levels of metabolisable energy and crude protein. In broilers in particular very high energy diets are produced, often containing significant quantities of fats or oils. However in most cases less than 50% of the ingested energy and nitrogen is retained in the animal carcass. This is illustrated in Table 6 where broilers retained only 38.8 % of energy and 48.2 % of protein (Aletor et al., 2000). This also means that over 50% of ingested dietary energy and nitrogen is excreted in breath, urine and faeces. There is clearly a pressing need to improve both energy and nitrogen utilization in food animals. This will bring environmental benefits as discussed in Chapter 9 as well as economic benefits in terms of production costs.

At the present time feeds are formulated to various nutrient values at the least cost. Performance of the feeds and efficiency of animal production is then usually judged by feed conversion ratio (FCR) and

Table 6
Energy and
nitrogen retention
by broilers from a
conventional
maize/soya feed

Major feed ingredients		Nutrients and nutrient utilization	
Maize (%)	49.4	Metabolisable energy (MJ/kg)	13.0
Soyabean meal (%)	40.4	Crude protein (%)	22.5
Soya oil (%)	6.2	Energy retained (%)	38.8
DCP (%)	1.8	Protein retained (%)	48.2

body weight. In Total Nutrition it would be more desirable to assess the combined biological and environmental efficiency of feeds. This would involve parameters as shown below:

BIOLOGICAL AND ENVIRONMENTAL EFFICIENCIES

Apparent energy retention (%):

$$\frac{\text{Gross energy consumed} - \text{Gross energy excreted}}{\text{Gross energy consumed}} \times 100$$

Apparent nitrogen retention (%):

$$\frac{\text{Grams nitrogen consumed} - \text{Grams nitrogen excreted}}{\text{Grams nitrogen consumed}} \times 100$$

The apparent energy retention and apparent nitrogen retention are simplistic but may have value in assessing Total Nutrition because of their simplicity. Feed energy is a difficult concept in that various levels of energy in feeds are recognised starting from gross energy to digestible energy to metabolisable energy to net energy. Feed nitrogen utilization is confounded with endogenous nitrogen losses. Nevertheless from a combined biological and environmental viewpoint the major factor is actually how much energy or nitrogen is consumed in feed and how much is either retained by the animal or returned to the environment in breath, faeces or urine. A major objective in Total Nutrition must be to maximise the amounts of energy and nitrogen retained by the animal. This must take into account levels of metabolisable energy and digestible nitrogen but this will be accounted for in the conventional feed formulations.

One approach to improving nitrogen utilization in livestock is by manipulating feed formulations to feed low-protein, amino acid-supplemented diets. The basic principle is to ensure more efficient nitrogen utilization by supplementing low-protein diets with essential amino acids.

A reduction in protein content of broiler feed from 22.5% to 15.3% did not affect the growth of the broilers when the feeds were supplemented with essential amino acids (Aletor et al., 2000). However as shown in Table 7 low protein diets resulted in increased feed consumption and consequently an increase in feed conversion ratio (FCR). A positive result was that both energy and protein retention increased in the low protein diets.

Table 7
Effect of low dietary protein supplemented with essential amino acids on broiler performance and nutrient utilization

Broiler performance	Feed formulation	
characteristic	Control (CP. 22.5%)	Low protein (CP 15.3%)
Feed intake (g)	2721	2960
Weight gain (g)	1522	1539
FCR	1.80	1.92
Energy retention (%)	38.8	46.0
Protein retention (%)	48.2	61.5

In broilers this strategy of feeding diets with low protein content together with amino acid supplementation has given variable results. In some cases this has led to impaired weight gain and feed efficiency whilst in other cases reduced protein formulations have performed identically to conventional formulations. The most consistent effect has been an increased deposition of abdominal fat in broilers fed low-protein diets. This is an undesirable characteristic from the consumers' point of view. Manipulation of feed formulations must encompass both optimal economic production and maintain attractive carcass quality and composition.

Future directions

The assessment of animal health and productivity has never been a simple process. However in future, systems of animal production where there will be less use of various drugs and medicines, will require more intensive surveillance of animals during their growth and productive periods in order to avoid diseases as well as maintain performance. As indicated above there are many physiological indices relating nutrition to health but many require expensive and sophisticated analytical procedures. This will clearly limit their application in intensive large-scale animal production. However there are some possibilities based on characteristics of the gut microflora, of blood, and of faecal matter, which might be of use to monitor Total Nutrition.

Gut microflora

It is widely accepted that the composition of the microflora of the gastrointestinal tract is important in animal health. In general beneficial bacteria such as the *Lactobacilli* need to be encouraged and pathogens such as *E. coli* or *Salmonella* discouraged. However, assessment of the gut microflora is difficult due to the diversity of the microbial populations and a technical inability to cultivate many of the viable bacteria found in the gastrointestinal tract of an animal. This makes it practically impossible to obtain a comprehensive picture of the activities of the microbial population in the gastrointestinal tract.

Recent advances in molecular biology have now led to the development of culture-independent methods for studying complex microbial eco-systems (Giraffa and Neviani, 2001). These techniques at present are highly sophisticated requiring specialised equipment and skills and so are not readily applicable to animal nutrition. Nevertheless in the future these techniques may play a useful role in helping to relate microbial ecology of the gastrointestinal tract to health and nutrition.

Blood

Blood is a readily available physiological material and may be used to evaluate the nutritional and health status of the target animals. The chemical composition of blood is under homeostatic control. Generally concentrations of various metabolites in blood are maintained within narrow limits but adverse conditions can influence this equilibrium. In reality blood plasma is more likely to be used for various clinical analyses. Blood plasma contains many different substances including lipoproteins, albumins, globulins and glucose, together with antioxidants such as ascorbic acid, uric acid, α-tocopherol, glutathione, zinc and carotenoids.

Many automated systems and test kits are already available for analysing blood plasma so these could readily be applied to animal health and nutrition studies. Literature on blood parameters of food animals in relation to health and nutrition is very scarce but is now receiving some attention. Turkeys for example raised under a high temperature regime had lower hematocrit values compared to those under a low temperature regime (Veldkamp *et al.*, 2000). Human subjects given green tea as a source of antioxidants showed an increase in the total antioxidant capacity of the plasma (Sung *et al.*, 2000). This work was conducted using a Total Antioxidant Status kit with automated clinical analysis and this would be technically feasible also for animal studies. In another study, blood plasma antioxidants in rats given carbon tetrachloride to generate an oxidative stress did not show any dramatic reduction (Kadiiska *et al.*, 2000). However administration of carbon tetrachloride causes liver malfunction and is far removed from nutritional studies. Nevertheless these examples indicate a possible approach that could be used in nutritional research in animal production.

Blood metabolite concentrations have been investigated in sheep as an indication of nutritional status (O'Doherty and Crosby, 1998). They investigated levels of ß-hydroxybutyrate, glucose, albumin, total protein, globulin and urea in the blood plasma of late pregnant ewes in relation to intake of protein and metabolisable energy. Plasma albumin and urea concentrations were significantly influenced by addition of soyabean meal to the diet.

Assay of acute phase proteins in blood is probably one of the most powerful tools available to objectively measure the health status of animals. As the technology for acute phase protein determinations becomes more widely appreciated it will play a major role in the assessment of animal health to optimise production rates. Acute phase proteins will provide information on the health of animals at slaughter, which will have implications for food safety. They can also be used to detect mastitis in dairy cows (Hirvonen, *et al.,* 1996).

Faeces

A significant amount of work has been done in ruminant nutrition using an *in vitro* gas production technique with faecal inoculations into feed mixtures (Theodorou *et al.,* 1994). This technique has also been applied to pig nutrition where faecal material was used to ferment cell wall fractions of different feed ingredients (van Laar *et al.,* 2000). The technique is quite attractive as the inherent fermentability of raw materials and the production of volatile fatty acids in the digested materials could be studied. Future developments will probably see more applications of this technique in other species although it was originally developed for ruminant studies.

Certain groups of micro-organisms in the gastrointestinal microflora, in particular the lactic acid bacteria or *Lactobacilli* are known to be effective in the inhibition of pathogens. Consequently an estimate of the potential efficacy of the population in the gastrointestinal tract to resist pathogens could be based on the proportions of *Lactobacilli*:coliform bacteria in the gut contents. This could perhaps be more easily followed by studying the condition of the faeces. In general a relatively large population of *Lactobacilli* in relation to coliform bacteria in the faeces would suggest a predominance of those bacteria inhibitory to pathogens and this might be a useful indicator of animal health.

References

Aletor, V. A., Hamid, I., Niess, E. and Pfeffer, E. (2000). Low-protein amino acid-supplemented diets in broiler chickens: effects on performance, carcass characteristics, whole-body composition and efficiencies of nutrient utilization. *Journal of the Science of Food and Agriculture*, **80**: 547-554.

Beck, M. A. (1999). Selenium and host defences towards viruses. *Proceedings of the Nutrition Society*, **58**: 707-711.

Bedford, M.R. (2000). Exogenous enzymes in monogastric nutrition – their current value and future benefits. *Animal Feed Science and Technology*, **86**: 1-13.

Benzie, I. F. F. and Strain, J. J. (1996). The ferric reducing ability of plasma (FRAP) as a measure of "antioxidant power": The FRAP assay. *Analytical Biochemistry*, **239**: 70-76.

Bovee-Oudenhoven, I. M. J., Termont, D. S. M. L., Weerkamp, A. H., Faassen-Peters, M. A. W. and Van der Meer, R. (1997). Dietary calcium inhibits intestinal colonization and translocation of salmonella in rats. *Gastroenterology*, **113**: 550-557.

Bovee-Oudenhoven, I. M., Wissink, M. L., Wouters, J. T. and Van der Meer, R. (1999). Dietary calcium phosphate stimulates intestinal lactobacilli and decreases the severity of a salmonella infection in rats. *Journal of Nutrition*, **129**: 607-612.

Chamanza, R., van Veen, L., Tivapasi, M. T. and Toussaint, M. J. M. (1999). Acute phase proteins in the domestic fowl. *World's Poultry Science Journal*, **55**: 61-71.

Diaz-Cruz, A., Nava, C., Villanueva, R., Serret, M., Guinzberg, R. and Pina, E. (1996). Hepatic and cardiac oxidative stress and other metabolic changes in broilers with ascites syndrome. *Poultry Science*, **75**: 900-903.

Dillard, C. J., Dumelin, E. E. and Tappel, A. L. (1977). Effect of dietary vitamin E on expiration of pentane and ethane by the rat. *Lipids*, **12**: 109-114.

Enkvetchakul, B., Bottje, W., Anthony, N., Moore, R. and Huff, W. (1993). Compromised antioxidant status associated with ascites in broilers. *Poultry Science*, **72**: 2272-2280.

Fairbairn, D. W., Olive, P.L., and O'Neill, K. L. (1995). The Comet Assay: A comprehensive review. *Mutation Research*, **339**: 37-59.

Giraffa, G. and Neviani, E. (2001). DNA-based, culture-independent strategies for evaluating microbial communities in food-associated ecosystems. *International Journal of Food Microbiology*, **67**: 19-34.

Hirvonen, J., Pyorala, S. and Jousimies-Somer, H. (1996). Acute phase response in heifers with experimentally induced mastitis. *Journal of Dairy Research*, **63**: 351-360.

Jackman, L. A., Millane, S. S., Martin, B. R., Wood, O. B., McCabe, G. P., Peacock, M. and Weaver, C. M. (1997). Calcium retention in relation to calcium intake and postmenarcheal age in adolescent females. *American Journal of Clinical Nutrition*, **66**: 327-333.

Kadiiska, M. B., Gladen, B. C., Baird, D. B., Dikalova, A. E., Sohal, R. S., Hatch, G. E., Jones, D. P., Mason, R. P. and Barrett, J. C. (2000). Biomarkers of oxidative stress study: Are plasma antioxidants markers of CCl_4 poisoning? *Free Radical Biology and Medicine*, **26**: 838-845.

Kneepens, C., M., F., Lepage, G., and Roy, C. C. (1994). The potential of the hydrocarbon breath test as a measure of lipid peroxidation. *Free Radicals in Biology and Medicine*, **17**: 127-160.

Lauridsen, C., Jensen, C., Jakobsen, K., Engberg, R. M., Andersen, J. O., Jensen, S. K., Sorensen, P., Henckel, P., Skibsted, L. H. and Bertelsen, G. (1996). The influence of vitamin C on the antioxidative status of chickens *in vivo*, at slaughter and on the oxidative stability of broiler meat products. *Acta Agriculturae Scandinavica, Section A, Animal Science*, **46**: 1-10.

Lauridsen, C., Hojsgaard, S. and Sorensen, M. T. (1999). Influence of dietary rapeseed oil, vitamin E, and copper on the performance and the antioxidative and oxidative status of pigs. *Journal of Animal Science*, **77**: 906-916.

Lawson, J. A., Rokach, J. and Fitzgerald, G. A. (1999). Isoprostane formation, analysis and use as indices of lipid peroxidation *in vivo*. *Journal of Biological Chemistry*, **274**: 2441-2444.

Loft, S., Fischer-Nielsen, A. and Jeding I. B. (1993). 8-Hydroxyguanosine as a urinary marker of oxidative DNA damage. *Journal of Toxicology and Environmental Health*, **40**: 391-404.

Lönnroth, I., Martinsson, K. and Lange, S. (1988). Evidence of protection against diarrhoea in suckling piglets by a hormone-like protein in sow's milk. *Journal of Veterinary Medicine B*, **35**: 628-635.

Morrow, J. D. and Roberts, L. J. (2000). The isoprostanes-unique bioactive products of lipid peroxidation. *Progress in Lipid Research*, **36**: 1-21.

New, S. A., Robins, S. P., Campbell, M. K., Martin, J. P., Garton, M. J., Bolton-Smith, C., Grubb, D. A., Lee, S. J. and Reid, D. M. (2000). Dietary influences on bone mass and bone metabolism: further evidence of a positive link between fruit and vegetable consumption and bone health. *American Journal of Clinical Nutrition*, **71**: 142-151.

O'Doherty, J. V. and Crosby, T. F. (1998). Blood metabolite concentrations in late pregnant ewes as indicators of nutritional stress. *Animal Science*, **66**: 675-683.

Plazier, J. C., Walton, J. P., Martin, A., Duffield, T., Bagg, R., Dick, P. and McBride, B. W. (2000). Effects of monensin on 3-methylhisitidine excretion in transition dairy cows. *Journal of Dairy Science*, **83**: 2810-2812.

Powell, S. R. (2000). The antioxidant properties of zinc. *Journal of Nutrition*, **130**: 1447S-1454S.

Qureshi, M. A., Hussain, I. and Heggen, C. L. (1998). Understanding immunology in disease prevention and control. *Poultry Science*, **77**: 1126-1129.

Rice-Evans, C. A., Miller, N. J., Boldwell, G. P., Bramley, P. M., and Pridham, J. B. (1995). The relevant antioxidant activities of plant-derived polyphenolic flavonoids. *Free Radical Research*, **22**: 375-383.

Rice-Evans, C.A. and Miller, N J. (1996). Antioxidant activities of flavonoids as bioactive components of food. *Biochemical Society Transactions*, **24**: 790-795.

Roberts, L .J. and Morrow, J. D. (1997). The generation and actions of isoprostanes. *Biochimica and Biophysica Acta*, **1345**: 121-135.

Robins, S. P. and New, S. A. (1997). Markers of bone turnover in relation to bone health. *Proceedings of the Nutrition Society*, **56**: 903-914.

Salguerio, M. J., Zubillaga, M., Lysionek, A., Sarabia, M. I., Caro, R., De Paoli, T., Hager, H., Weil, R. and Boccio, J. (2000). Zinc as an essential micronutrient: a review. *Nutrition Research*, **20**: 737-755.

Singh, B. and Paraskeva, C. (1998). Bcl-2 as a possible sensor of nutritional stress inhibiting apoptosis and allowing cell survival during colorectal carcinogenesis. *Biochemical Society Transactions*, 664th Meeting University of Reading, **26**: 236-241.

Strain, J. J. (1999). Optimal nutrition: an overview. *Proceedings of the Nutrition Society*, 395-396.

Sung, H., Nah, J., Chun, S., Yang, S. E. and Min, W. K. (2000). *In vivo* antioxidant effect of green tea. *European Journal of Clinical Nutrition*, **54**: 527-529.

Surai, P. F. and Speake, B. K. (1998). Distribution of carotenoids from the yolk to the tissues of the chick embryo. *Journal of Nutritional Biochemistry*, **9**: 645-651.

Theodorou, M. K., Williams, B. A., Dhanoa, M. S., McAllan, A. B. and France, L. (1994). A simple gas production method using a pressure transducer to determine the fermentation kinetics of ruminant feeds. *Animal Feed Science and Technology*, **48**: 185-197.

Van den Berg, R., Haenen, G. R. M. M., van den Berg, H. and Aalt, B. (1999). Applicability of an improved trolox equivalent antioxidant capacity (TEAC) assay for evaluation of antioxidant capacity measurements of mixtures. *Food Chemistry*, **66**: 511-517.

Van Laar, H., Tamminga, S., Williams, B. A., and Verstegen, W. A. (2000). Fermentation of the endosperm cell walls of monocotyledon and dicotyledon plant species by faecal microbes from pigs. The relationship between cell wall characteristics and fermentability. *Animal Feed Science and Technology*, **88**: 13-30.

Veldkamp, T., Kwakkel, R. P., Ferket, P. R., Simons, P. C. M., Noordhuizen, J. P. T. M. and Pijpers, A. (2000). Effects of ambient temperature, arginine-to-lysine ratio, and electrolyte balance on performance, carcass, and blood parameters in commercial male turkeys. *Poultry Science*, **79**: 1608-1616.

Wood, L. G., Fitzgerald, D. A., Gibson, P. G., Cooper, D. M. and Garg, M. L. (2000). Lipid peroxidation as determined by plasma isoprostanes is related to disease severity in mild asthma. *Lipids*, **35**: 967-974.

Wu, G. Y., Field, C. J. and Marliss, E. B. (1991). Elevated glucose metabolism in splenocytes from spontaneously diabetic BB rats. *Biochemical Journal*, **274**: 40-54.

9   Difficult Demands: Safe Food, Low Cost, Ethical Issues, Environmental Impact

The raising of animals for food increasingly has to satisfy critical and sceptical consumers, and modern consumer demands frequently appear somewhat incompatible. For example large volumes of cheap food are desired which are absolutely safe, come from animals maintained under good welfare conditions and has no adverse impacts upon the environment. Total Nutrition is an attempt to respond to these demands by a nutritionally focused programme which encompasses animal health, welfare and productivity. Many of the facets of Total Nutrition discussed in the previous chapters are however also quite relevant to human nutrition and health. There is increasing interest and concern today in the connections between health and nutrition both for animals and humans. It is important to be able to demonstrate to the sceptical consumer that animal nutrition is an integral part of human nutrition and that standards in animal and human nutrition are moving closer together.

The enormous increase in the productivity of animals raised for food is of great benefit to the consumer as this is the foundation of the widespread availability of large volumes of low cost food. It is no mean feat that even as populations have increased food supplies has more than kept up with this population growth in the developed countries. A comparison between animal production characteristics in the USA in 1930 and today is illustrative of this important point. In 1930 US milk production was about 2200 litres per cow per year. Over the last 70 years milk production per cow has jumped four-fold to over 8000 litres a year. Due to advances in pig nutrition the market age for pigs has dropped from 200 days in 1930 to 120 days, a reduction of 40%. Beef cattle are ready for market in half the time it took in 1930. A laying hen produces four times as many eggs in a year as it could in 1930. It now takes 60% less feed to produce 1 kg of chicken in the US than it did 70 years ago (Bossman, 2001).

This improvement in productivity of animal husbandry has enormous implications for society and for the environment. If productivity had remained at 1930 levels in the USA they would need four times as many dairy cows and four times as many laying hens as are actually raised today to provide the same amount of food for the population. To produce the quantity of milk, eggs, broilers and pigs required today with 1930s production levels would require a huge increase in amount of land and of feed required with a concomitant increase in environmental pollution. This is clearly undesirable and continuous advances in productivity need to be made to support population growth and to minimise the environmental impact.

It is also important to emphasise that animal production is not an isolated system or activity, but is an integral part of modern society. Furthermore it is an activity which has significant beneficial economic and environmental effects in the re-cycling of low-value food materials which in many cases cannot be efficiently utilised by humans. If these low-value materials cannot be re-cycled through animal nutrition then disposal becomes another problem. Alternative disposal systems such as incineration, land filling or dumping into the seas are also in many cases unsatisfactory alternatives to re-cycling through animal nutrition.

The great sociological value of being able to produce large quantities of low cost food of animal origin should not be overlooked in all the rhetoric and media attention on animal agriculture. It would be difficult to argue that food should be high-cost and consequently widely available only to the richest section of the population. The current situation where much adverse publicity is directed towards animal production is to some extent a problem of the success of the animal production industry such that in the developed world access to unlimited quantities of cheap food of animal origin is taken for granted.

In the developing world by contrast increased meat consumption is seen as a feature of economic development and as an entitlement by most populations. As prosperity increases *per capita* consumption of staple cereals such as rice declines and is replaced by foods of animal origin; meat, milk and eggs. For example in Japan in the early 1960s *per capita* rice consumption was 120 kg but now has declined to 68 kg *per capita* (Khush, 2001). Increased demands for meat consumption with increased affluence is an inevitable part of economic advancement.

There is at present an enormous difference in meat consumption in the developed world compared to the developing world (Rosegrant *et al.*, 1999) (Table 1). In the developed countries annual *per capita* meat consumption over the years 1981-94 was 76 kg, compared with only 17 kg for a person in the developing countries. Recent projections have shown that world meat demand will grow at a rate of 1.8% per year from 1993-2020.

Table 1
Total and *per capita* meat consumption over the years 1961-1994

Region	Meat consumption	1961-71	1971-81	1981-94
Developed countries	Total meat	59.66	79.24	94.20
Developing countries	Consumption	23.96	36.07	65.77
World	(million tonnes/year)	83.62	115.31	159.96
Developed countries	*Per capita* meat	57.61	69.93	76.46
Developing countries	Consumption	10.05	11.93	17.01
World	(kg/year)	24.47	27.82	31.55

It can be estimated from Table 1 that if annual *per capita* meat consumption in the developing countries were to rise to the same level as that currently in the developed countries then total world meat production would need to double from about 160 million tonnes to 320 million tonnes.

Certainly when economic conditions improve low-income populations tend to increase consumption of meat and this usually brings with it reduced incidences of nutritional deficiencies. Increased meat consumption in the developing world can only be viewed as a desirable outcome of improved economic conditions. However it does pose serious questions of availability of raw materials to produce the feed necessary to support the numbers of animals required.

Nevertheless within the modern system for the production of large volumes of low-cost food there are several aspects of food production of importance to the consumer that Total Nutrition needs to address. These are issues of food safety, utilization of feed resources, animal health and welfare and environmental impact. It is important to demonstrate to the modern consumer and to the political authorities, that the system of production of foods of animal origin, as outlined in Total Nutrition, is an integral part of a sustainable system. Furthermore this system as currently used is highly developed, science-based, and is an intelligent use of natural resources. The alternative in reality would be sustainable poverty for the majority of the population.

Food safety

Obviously food for the human consumer must be absolutely safe. Recent food safety issues of *Salmonella,* BSE and dioxins naturally raise fears in consumers concerning the safety of their food. These issues have been related to animal feed quality and it is obvious that feed quality has an important impact upon food safety. Feeds must be free from pathogenic micro-organisms and from toxic materials, particularly mycotoxins.

The strategies described in Chapter 2 illustrate that the knowledge and technology to produce safe animal feed are widely available. If these are rigorously followed then both feed raw materials and animal feeds can be produced which are safe and of good nutritional quality.

Modern animal feed production however must not only be safe but the safety must be demonstrated. This can be achieved by the implementation of various quality assurance systems such as GMP (Good Manufacturing Practice) HACCP (Hazard Analysis at Critical Control Points) and various ISO 9000 standards.

There is also significant concern about zoonotic diseases spreading throughout the food chain. Zoonoses are infections that are transmitted from animals to man and at the present there are seven food-borne zoonoses of concern for public health. These are diseases caused by *Salmonella, Listeria monocytogenes, Campylobacter*, Verotoxigenic *Escherichia coli, Cryptosporidium, Echinococcus granulosus* and *Trichinella spiralis.*

In the EU there is now considerable emphasis upon reduction of *Salmonella* in pigs and of *Campylobacter* in poultry. In humans both organisms cause similar food poisoning symptoms of headache, fever, vomiting and diarrhoea. Fortunately illnesses caused by *Campylobacter* tend to be less severe than those caused by *Salmonella* and rarely last longer than one week.

These two major problems, *Campylobacter* in broilers and *Salmonella* in pigs are somewhat similar in that the organisms of concern are not pathogens for the host animal. This has some important implications for control of these organisms in animal production. It will be difficult if not impossible to show any growth or performance benefit from control of these organisms in the host animal. Therefore programmes requiring treatment of the flock or herd for the whole life cycle will be difficult to justify economically. The interest of the producer will be to have broilers or pigs free from the pathogen at time of slaughter. This suggests that programmes will need to be developed that can 'clean' the animals in the late stages of the growth period.

Organic acid nutricines might play a useful role here (Byrd *et al.*, 2001). Addition of 0.5% lactic acid to the drinking water of broilers over a 10 hour period when feed was withdrawn significantly reduced crop contamination with *Campylobacter.*

Control of zoonoses is acknowledged to be a difficult task as the causative organisms are ubiquitous. This makes it difficult to eliminate them completely from the food chain. Epidemiological data collected in the EU reveals that food-borne zoonotic diseases have increased over the last 20 years. There is the intention within the EU to establish a system of monitoring and control of the entire food chain from the farm to the fork and some aspects of Total Nutrition must address control of zoonoses.

Zoonotic diseases are becoming increasingly important in human health as there are several examples of viral diseases which can jump from animals to humans. Perhaps the most well know example is HIV, the causative agent of the aids syndrome. It is widely accepted that HIV jumped from apes to humans sometime ago possibly as early as the 1930s. It appears that many viruses may be tolerated by various animal species, such as HIV in apes, yet after crossing over into another species this same virus becomes a deadly disease. Several other viruses in recent years have been observed to infect humans often with fatal consequences. Arena virus which includes Lassa fever, Ebola virus, Ross river virus, Hanta virus, and Lyme disease have all been reported as infecting humans. These are not food or feed-borne so nutrition is not directly implicated in these diseases. Nevertheless there is a perennial risk to health of humans and of their animals posed by an enormous array of virulent micro-organisms. They have evolved a powerful ability to mutate and to generate genes for virulence and for drug-resistance. This makes it even more important that animal and human nutrition is directed towards health maintenance, with support of the immune system in order to ensure that humans receive highly nutritious food which in turn should help them to withstand various zoonotic and other diseases.

The shelf life of foods of animal origin is also very important as little food in the modern world is consumed at the point of production. Most of our food is produced at a location remote from the point of purchase and consumption. This inevitably means that foods must be stored for a period of time and transported considerable distances before they are actually eaten by the human consumer. The maintenance of good organoleptic characteristics and the absence of pathogens is very important in the concept of "shelf-life." It is widely recognised for example, that a good antioxidant supply in the feed of animals raised for food gives meat products with improved shelf-life. Milk and eggs can be carriers of pathogenic bacteria and so it is important that these dangers are controlled through Total Nutrition.

Antibiotic-free food

Consumers want safe, cheap food products of animal origin. They also want these produced without recourse to widespread use of antibiotics and other medicines in food animals. As discussed in Chapter 1 there is widespread concern about the use of drugs and medicines in animal production in terms of resistance of pathogenic bacteria to antibiotics. One of the major objectives in Total Nutrition is to establish systems of animal production which do not rely upon widespread use of antibiotics and other drugs. This requires a detailed understanding of the roles and functions of various nutrients and nutricines. However it should be possible to meet this requirement by careful feed formulation and good standards of animal husbandry. This requires considerable attention to animal health issues and feeding for health maintenance as well as for growth is now very important. Total Nutrition will be of great value to the consumer if it allows the production of food of animal origin without recourse to antibiotics or other drugs.

Ethical issues: utilization of food resources, efficiency of animal production and animal welfare

There has been considerable debate in recent years over the moral and ethical issues of using grain or cereals to feed animals rather than for direct nutrition of humans. It is often claimed that feeding cereals to animals and then eating the animals is much less efficient than eating the cereals directly (Millward, 1999). Frequently gross calculations about animal production and grain use are made. For example production figures have been presented to show that it requires 2 kg, 4 kg and 6 kg of grain to produce 1 kg of poultry, pork and beef respectively (Khush, 2001). The implication is that this is an inefficient use of scarce food materials. However it is a rather simplistic analysis and takes little account of the actual forms and types of grains used and of the mechanisms of animal nutrition and production using modern techniques.

All animal and human nutrition ultimately depends upon only five basic types of raw materials, all of plant origin; forages, oilseeds, cereals, fruits and vegetables. These five classes of materials must feed both humans and animals. A very large proportion of the five basic classes of food raw materials cannot be utilised directly by humans especially forages, and many other feed ingredients are actually inedible by-products of human food manufacturing. Humans as omnivores require foods of both animal and plant origin for optimum growth and development.

Forages, which include pasture, silages, hay and straw, are produced in very large quantities and cannot be utilized by humans at all in their native state. Consequently ruminants play an important role in the ecosystem as they can digest these high fibre forages which humans cannot utilise and convert them into valuable human food products such as meat and milk. Ruminants are also an important source of non-food items such as leather and wool.

It should also be emphasised that many animal feeds for non-ruminants such as pigs and poultry actually contain a large amount of raw materials which are basically inedible for humans. The judicious use of least cost feed formulation, application of appropriate nutricines, improved genetics and high quality stockmanship allows ever-greater use of raw materials in feeds which are either inedible or poorly edible for humans. These low-grade raw materials are then converted into high-value human food products by feeding them to animals. The manufacture of animal feed and the raising of animals for food serves a very important function in converting materials which cannot be directly digested by humans into valuable sources of meat, milk and eggs.

Modern ruminant production also relies less and less on feeding animals human food-grade cereals. For example dairy rations can be, and frequently are, produced without any cereal components whatsoever. Dairy cows can be adequately fed upon a forage source, usually silage or fresh grass together with a manufactured concentrate feed that does not necessarily contain any human food-grade cereals.

It is also important to point out that arguments such as it takes 6 kg of grain to produce 1 kg of beef (Khush, 2001), are mostly relevant only to a particular area, the USA. This is not necessarily competing with humans for food resources but reflects to a large extent the internal economics of the USA where large quantities of cereals, particularly maize, can be produced at low cost and efficiently transported and stored. In this scenario feeding maize to beef cattle is economically justified but it is not an essential requirement for the raising of beef cattle. In other regions of the world where other economic or social conditions prevail modern animal production systems can dispense with grain utilization for most ruminant production. For example in the EU and the Argentine beef cattle are not fed on large quantities of maize but utilize significant quantities of forages either as silage or pasture.

Modern animal production also plays a very important role in the recycling of waste products and by-products from the human food manufacturing industry. This has the dual advantage that poor quality raw materials are converted into high quality food products of animal origin and these materials are disposed of in an efficient manner and not left to accumulate in the environment. This aspect of modern animal production in the developed world is of major importance in avoiding excessive environmental pollution by waste products from the food manufacturing industries.

The classic soyabean meal so widely used in animal feed formulations is in fact a by-product from the edible oil industry. Soyabean oil is widely used in human foods and this generates in turn massive quantities of extracted soyabean meal. Only a relatively small proportion of this soyabean meal is further processed into a form suitable for direct human nutrition. The basic extracted soyabean meal is of little value for direct human nutrition, but is widely used in animal feeds. Similar situations occur with rapeseed, sunflower seeds and oil palm. Indeed, rapeseed meal, sunflower meal and palm kernel meal have even less value in direct human nutrition than has soyabean meal, but are nevertheless widely used in various animal feed formulations.

Production of sugar from sugar beet generates enormous quantities of beet pulp. Processing of sugar cane generates bagasse. Neither of these materials is ever going to be consumed directly by humans and therefore they have no other outlet than in animal feed.

Similarly the wheat-, maize-, rice- and other cereal-milling industries which process grains into products for direct human consumption generate large volumes of inedible by-products such as wheat bran and middlings, maize gluten, and rice bran. Further processing of cereals in the fermentation of alcoholic beverages such as beer, wines and sprits also produce large amounts of brewers' grains and distillery by-products, which are efficiently re-cycled through animal feeds.

The fruit and vegetable processing industries generate large quantities of materials inedible for humans but of value in animal nutrition. By-products from the potato industry for example are widely used in animal feeds.

Consequently as illustrated in Figure 1 the choice in the utilization of these basic food and feed raw materials is not simple but extremely complex apart from forages which can only be used for animal feeding.

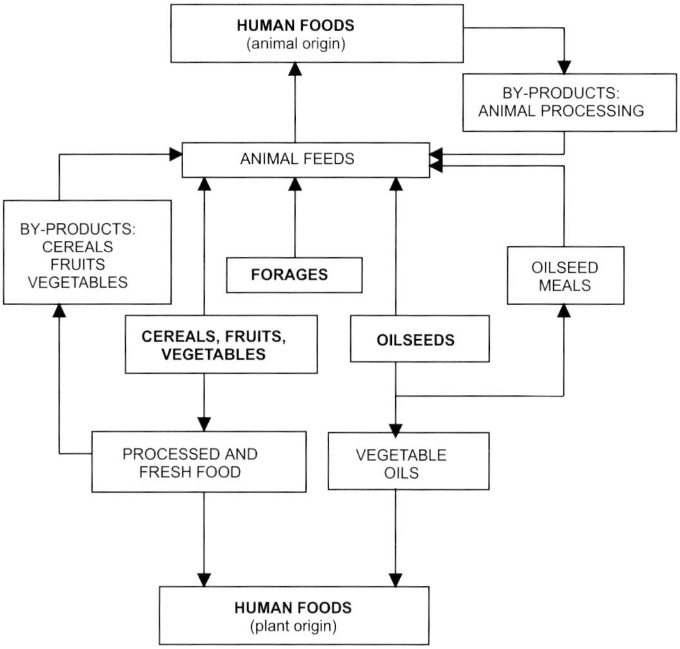

Figure 1
Conversion pathway of basic food raw materials into human foods of plant and of animal origin

For example in Germany compound feed manufacturers process some 19 million tonnes of raw materials each year, and they consist of just under 40% of primary products (cereals, legumes, and cassava). The remaining volume is made up almost exclusively of by-products from food manufacture. These low value by-products are utilised efficiently in animal production and also remove a potential environmental threat (Grote and Radewahn, 2000).

One of the major challenges in Total Nutrition is to achieve greater efficiencies in the conversion of basic raw materials and of by-products into high value products of animal origin for human nutrition. When the details of modern animal nutrition are considered it is clear that animal production is not only a consumer of cereals, but also an extremely efficient way of converting many low-grade basic raw materials into high value products.

Broiler chickens can be produced with a feed conversion ratio (FCR) of as low as 1.60. Broiler feed contains about 65% cereals so this means that 1 kg of broiler meat is obtained from 1.04 kg of cereals. This 65% cereal content is in fact milled whole cereals which would not be directly consumed as such by humans anyway. The other 35% of the broiler diet is made up of ingredients such as soyabean meal, vitamins and minerals none of which compete with human nutrition. Aquaculture converts 1 kg of fish feed into more than 1 kg liveweight of fish. Here the major ingredients are fishmeal and fish oils which again are not likely to be consumed directly by humans. In Europe most beef and dairy cows are fed a mixture of grass or silage and a manufactured feed which is frequently cereal-free and composed of various by-products from the human food industry. Indeed Total Nutrition, which makes effective use of both nutrients and nutricines in feed formulations, actually has a feed sparing effect and a positive environmental impact.

Feed sparing effect

The data in Table 1 illustrate that for citizens of the developing countries to increase their meat consumption to that of the developed countries, meat production must double. Such a great increase in meat production will require ever-greater efficiencies in feed utilization by the animals. In the developed countries over the last 70 years there has been enormous improvements in productivity of animals raised for food (Bossman, 2001). Furthermore an enormous number of materials can be used in animal feeding in addition to grains or cereals and many of these raw materials are not suitable for direct consumption by humans.

The concepts outlined in Total Nutrition are directed towards the feeding of animals for health and growth which will result in improved yields of food products of animal origin obtained from the same quantity of feed raw materials. These improvements in production efficiency through the application of Total Nutrition will be extremely important in allowing ever-greater volumes of food items to be produced from a given quantity of feed raw materials. Information gathered over the last few decades convincingly demonstrates that significant improvements in the efficiency of animal production have been made. The value of these improvements in the productive efficiency of animals can be easily demonstrated in terms of a feed sparing effect as illustrated by the calculations in Table 2 for broilers and pigs.

217

Growth characterisitic	Broilers[1]	Pigs[2]
Initial FCR	2.0	3.0
Feed required (kg/animal)	4.0	210.0
Improved FCR	1.8	2.5
Feed required (kg/animal)	3.6	175.0
Feed Sparing Effect (kg)	0.4	35.0

[1]Broilers grown to 2.0 kg; [2] Pigs grown from 30-100 kg.

The calculations shown in Table 2 illustrate that fairly modest and realisable improvements in animal productivity have very significant effects in terms of feed requirement. It is quite feasible under modern production systems to reduce the feed conversion ration of broilers from 2.0 to 1.8 or even lower. This means that the feed required to produce 1 kg of broiler will drop by 400 g and hence improved productive efficiency gives a feed sparing effect. When this feed sparing effect is multiplied up by the many millions of broilers or pigs produced per year this becomes a very significant quantity of feed that has been spared. For example production of 1 million broilers with an FCR of 2.0 will require 4,000 tonnes of feed. However if the same quantity of broilers can be produced with an FCR of 1.8 then only 3,600 tonnes of feed will be required. Alternatively for the same 4,000 tonnes of feed now 1,052,632 broilers could be produced.

For pigs the feed sparing effect is also very significant in that a production of 1 million pigs with an improved FCR of 2.5 would spare some 35,000 tonnes of feed. It is obviously extremely important to continuously improve the efficiency of animal production in order to feed an increasing population.

The data presented in Table 2 is a theoretical example to demonstrate the significance of Total Nutrition in overall global economics of feed raw material utilization in relation to the production of high quality human food items. These generalised data are however supported by practical experience as shown by data from laying hens collected over the years 1972-1996 (Table 3) (Flock, 1998).

Years	FCR (kg feed/kg egg mass	
	White egg layer	Brownegg layer
1972-76	2.69	2.86
1977-81	2.54	2.68
1982-86	2.46	2.48
1987-91	2.33	2.32
1992-96	2.30	2.18
Feed sparing effect (kg/kg egg mass)	0.39	0.68

Obviously there has been a very significant saving in quantities of feed needed to produce 1 kg of egg mass in both white and brown layers. Over the period 1972-1996 it has become possible to produce 1 kg of eggs with on average about 500 g less feed. Again when this is multiplied by the many millions of kg of eggs produced there is a very significant feed sparing effect as a result of modern developments in poultry science.

Broiler production shows similar significant feed sparing effects. For example, over the years 1989-96 the FCR of broilers dropped from 1.984 to 1.843 (Cobb, 1999). This is a feed sparing effect of 141 g per kg of broiler which again is an enormous saving in feed used in total broiler production. In addition body weights increased over the same period from 1.80 to 2.25 kg. This is an added advantage in that fewer birds would have been required in 1996 to supply the same quantity of meat compared to the number required in 1987.

The feed sparing effect has possibly greater potential in pig production than in layer or broiler production because of the unusual digestive physiology of the pig. In particular the physiology of the large intestine of the pig is similar in some characteristics to that in humans and also to that of ruminants. However there are also important differences. The pig large intestine contains all the major cellulose degrading bacteria that occur in the rumen but these are not found in humans. Unlike the rumen however, the pig large intestine does not contain protozoa and there is less methane production in pigs than in ruminants.

Adult pigs in particular have a great ability to utilize effectively high levels of dietary fibre. In this respect they are much more efficient than humans (Varel and Yen, 1997). Dietary fibre may contribute up to 30% of the maintenance energy needs of growing pigs. Even greater energy contributions can be obtained from fibre fed to sows and this can bring associated benefits in terms of health and reproductive efficiency. A portion of cereal grains in mature pig feeds can be replaced by fibrous ingredients which are unacceptable for direct human consumption. These include lignified plant material from forages such as grasses, straw and legumes. Various cereal milling and brewing or distillery by-products with a high fibre content can also be used. It is important to reiterate that all these materials cannot be directly consumed by humans. Therefore use of these in pig diets is an effective way to transform essentially inedible and even waste materials into valuable food products for human nutrition. This will reduce the amount of grain consumed and give a feed sparing effect.

Far from competing with humans for food resources or wasting human food resources as is so often suggested, modern animal production can ensure efficient transformation of vast quantities of materials inedible for human consumption. Further developments in animal nutrition will make ever-greater contributions to the feed sparing effect. This will ensure that the maximum amounts of high quality human food products are obtained from the available supplies of basic feed ingredients.

Animal welfare

Animal welfare is a highly emotive topic and is not easy to describe in hard scientific or practical terms. Most people would agree that animals raised for food should not be subjected to unnecessary hardship or cruelty. This implies that modern systems of animal production must pay due care to the welfare of the animals.

There have been several major issues of animal welfare in the EU in recent years such as the use of sow tethers, docking of piglet tails, the use of battery system for layers, feed restrictions for breeding poultry and gestating sows.

Concern has also been expressed over live animal transport, particularly veal calves which may be transported in lorries for several thousands of kilometres. Transport induced stress on broilers increased the shedding of *Campylobacter* species in faecal material and this may subsequently result in extensive carcass contamination (Whyte *et al.*, 2001). This is clearly undesirable from both the food safety and welfare points of view.

Welfare of an animal may be defined as its state when it attempts to cope with its environment and various indicators of welfare have been proposed as shown in Table 4 (Broom, 1996).

Table 4
Indicators of welfare, reducing from good to poor

Variety of normal behaviour displayed
Physiological indicators of pleasure
Behavioural indicators of pleasure
Reduced life expectancy
Reduced ability to grow or breed
Extent of body damage
Extent of disease
Extent of immunosuppression
Extent of behaviour pathology
Extent to which normal physiological processes and anatomical development are prevented

Those parameters of behaviour which have a physiological base such as physiological indicators of pleasure and extent of immunosuppression cannot easily be monitored in herds or flocks. Other aspects of general welfare, which are clearly correlated with good health, are easily monitored and are of interest in terms of efficient animal production as well as welfare. Food animals have generally been selected for good growth and reproductive performance over many generations. Clearly if growth and performance are below the expected values welfare is likely to be poor. This would frequently also lead to reduced financial returns which in itself would be an incentive to maintain good standards of welfare.

One physiological indicator of animal welfare may be to measure acute phase proteins. As described in Chapter 8 these are proteins found in blood plasma which increase in concentration in response to various stresses. Assay of the concentration of these proteins in blood can provide an objective measure of the health status of an animal and are increasingly being used as markers of both animal health and welfare.

The relationship between transport stress and acute phase proteins has been convincingly demonstrated in boars (Piñeiro *et al.*, 2001). Two groups of boars were transported long distances under different conditions. Group 1 had a journey of 48 hours under excellent conditions of $2m^2$ /boar, sawdust, feed and water provided. Group 2 had a journey of 24 hours under average transport conditions of 1.5 m^2/boar no sawdust, feed or water provided. Boars from Group 2 demonstrated significantly higher levels of acute phase proteins in the blood than in boars from Group 1 despite the journey lasting 24 hours longer in Group 1. Transport stress substantially affects concentration of acute phase proteins and these have a great potential as indicators of animal welfare.

Environmental impact

Large-scale intensive animal production has a very marked and frequently serious environmental impact. Manure production by the large number of animals needed to supply food is of major concern in many countries. This can be manifested as objectionable odours due to manure production or increased fly populations. In many cases there is a surplus of nitrogen and phosphorus in the manure applied to land which exceeds the nutritional requirements of plants. Excessive application of manure to land also results in the accumulation of heavy metals with potential risks for both plant growth and animal health.

This leads to more insidious effects such as pollution of water courses as a result of disposal of animal wastes. In soil most of the nitrogen from manure is transformed into nitrate. This is a form of nitrogen readily utilized by plants but is also very soluble and is easily leached out into water courses. Washing of undigested feed nutrients, particularly phosphorus into water, leads to eutrophication and loss of oxygen in the water.

Several solutions and programmes are being implemented to deal with the pollution problems of intensive animal production. There are various administrative procedures in force to prevent expansion of animal production in certain areas. Various financial penalties are levied on farms and units which produce manure. Most of these solutions are extremely costly and they do not focus on the origin of the problem which is the transformation of nutrients by the animal.

Many of these effects can be reduced or substantially alleviated through application of the strategies outline in Total Nutrition. Good management of animal feeds and feeding is a valuable tool for reducing nutrient excretion into the environment.

Environmental nutrient load (ENL)

The fundamental aspect of animal nutrition which causes concern from environmental pollution is the amount of various nutrients excreted by the animals into the surrounding environment. This is the environmental nutrient load (ENL) and it is influenced by several factors and is the final result of complex interactions of digestion, absorption, retention and excretion of nutrients by animals.

The ENL can be expressed in the form of a simple equation as:

$$ENL = (\text{Nutrient intake} - \text{Nutrients utilized}) + \text{Nutrients from endogenous losses} \qquad (1)$$

The first contribution to the ENL is clearly the nutrient intake or the amount of feed consumed. The feed sparing effect described above will have a beneficial impact upon the environment in that less feed needs to be consumed by a given animal as productive efficiency increases.

Only a part of the nutrients in the ingested feed is used for maintenance and productive growth of the animal. The remaining nutrients are excreted into the environment in the form of manure, urine or gasses. Nutrient utilization is itself a complex topic and is influenced by the factors already discussed in previous chapters. The interaction of various nutricines and nutrients has a major impact upon nutrient utilization, absorption and retention and therefore also upon the ENL. Improvements in the proportion of nutrients utilised will have beneficial effects both in terms of overall economics and upon the ENL.

Endogenous nutrient losses is another contribution to the ENL. In any living animal there are always some endogenous losses of cells, recycled nutrients and enzymes. However there is little that can be done at present to influence the amount of endogenous secretion.

In both ruminants and monogastric animals the relative supplies of energy and protein in the diet will have a great impact upon the ENL. In ruminants there must be sufficient fermentable energy available in the rumen to support microbial protein synthesis and to reduce the excretion of nitrogen as urea. In monogastric species reduced dietary proteins and better nitrogen utilization by the animals will reduce the ENL.

It is obvious from Equation (1) that the ENL will be reduced if nutrient intake and endogenous losses are decreased and if nutrient utilization is increased. The greatest impact on nutrient intake will be obtained by reducing the number of animals raised. This is where nutrient utilization and general production efficiency become so important both for food supply and pollution control. Highly efficient animal production allows food to be produced from fewer animals with less feed consumed. Total Nutrition with the objectives of efficient and productive feeding of animals to support growth and maintain health will also contribute to minimise environmental pollution.

Nitrogen excretion

In pigs and poultry for example, nitrogen retention by the animal is usually about 50% or less of the nitrogen consumed in feed. Growing pigs use only 30-35% of ingested dietary nitrogen and phosphorus (Jongbloed and Lenis, 1992). This means conversely that generally over 50% of the nitrogen consumed is excreted into the environment and in sows some 75% of nitrogen consumed is excreted (van der Peet-Schwering and den Hartog, 2000). This low efficiency of nitrogen utilisation makes intensive animal production a serious contributor to nitrogen pollution of the environment. Farm effluents from animal production systems are also a major source, nitrate pollution of water systems.

The amount of nitrogen excreted by animals is related to nitrogen intake in the diet and in modern nutrition there is often an excess of dietary nitrogen as a security measure. Proteins are considered in feed formulations as a supply of various amino acids. The requirement of any essential amino acid for any rate of growth has to be present in a fixed ration relative to the others in a diet. This is known as the "ideal protein" concept. Unfortunately many feed formulations cannot match the "ideal protein" and supply an excessive quantity of non-essential amino acids in order to ensure an adequate supply of the essential amino acids. There has been a considerable amount of effort expended by various research organizations to develop the ideal protein concept (Ketels, 1999).

Feeding diets with a reduced nitrogen content does not necessarily reduce the performance of growing pigs if the amino acid balance is well managed (Tuitoek *et al.*, 1997). Reducing dietary crude protein content in feed from 16.6 % to 13.0% for pigs from 20-55 kg had no significant effect upon average daily gain, feed conversion ration or protein deposition.

However reducing crude protein levels in feed will reduce the amount of nitrogen excreted as illustrated in Table 5 with data on growing and finishing boars (Lee, 2001).

In this case, nitrogen excretion (g/day) was reduced by using feeds with lower crude protein levels, but also absolute nitrogen retention (g/day) by the animals was reduced, and this will have an undesirable effect upon performance. It is interesting to note that the proportion of nitrogen retained as a percentage of nitrogen intake was not so strongly affected by reduced crude protein content of diets and averaged 57%. This implies that nitrogen utilisation was not greatly changed by reducing crude protein levels in the feed. The real challenge of course is to improve nitrogen utilisation from the feed so that animals can retain a higher proportion of the nitrogen consumed and concomitantly there will be less nitrogen to contribute to the ENL.

Table 5
Absolute nitrogen retention (NR, g/day) and nitrogen retention as a percentage of nitrogen intake (NR, %)

Growing boars	Crude protein (%)	25.0	21.0	18.0
	Nitrogen excreted (g/day)	30.2	24.2	17.7
	Absolute NR (g/day)	39.3	35.0	32.3
	NR (% nitrogen intake)	57	59	65
Finishing boars	Crude protein (%)	22.0	18.5	16.0
	Nitrogen excreted (g/day)	39.5	30.2	27.0
	Absolute NR (g/day)	42.3	39.5	32.5
	NR (% nitrogen intake)	51	57	55

The general strategy is to feed animals with the appropriate levels of essential amino acids for optimal performance while limiting the excess of protein that is ingested. Formulation of low protein diets consists of reducing the incorporation of protein-rich feedstuffs such as soyabean meal and balancing the diets with supplementary amino acids. These would be lysine, methionine threonine and tryptophan, although other essential amino acids might be introduced in the future. This should ensure that as much as possible of the ingested feed is actually digested and absorbed by the animal rather than voided as manure back into the environment. Application of nutrines such as enzymes and phospholipids are particularly important here. More efficient feed utilization due to enzyme nutricines will allow reductions in nitrogen content of feeds and consequently less nitrogen will be excreted into the environment.

Decreasing the protein level of feeds by 1% in general results in a 10% reduction of nitrogen excretion in both pigs and poultry production. It should therefore be possible with current feed ingredients and amino acids to achieve a total reduction of 25% of nitrogen excreted.

Animals fed low protein, amino acid supplemented diets have less excess amino acids in the diet that need to be catabolised and excreted in the urine as uric acid. This should lead to a reduced production of urine, and a decrease in water intake. This will bring further benefits such as a sparing of water, higher dry matter content of the slurry, and a decreased volume of manure to be transported and distributed.

It seems very possible to reduce protein contents of pig diets for environmental reasons and maintain good production. In broilers however it is less feasible and evidence is somewhat contradictory. A fairly consistent observation is that low protein diets in broilers give increased abdominal fat deposition and this is undesirable. Fat retention in broilers increased from 146 g to 289 g as protein content of the feed dropped from 22.5% to 15.3% (Aletor *et al.*, 2000). This represented a two-fold increase in fat deposition. Surprisingly, equalising the dietary crude protein levels to that of the control feed with supplementary non-essential amino acids in order to maintain identical protein:energy ratio did not correct the increased fat deposition. The low protein diet with 15.3% crude protein did reduce nitrogen excretion by some 41% compared to the control feed so this is a consistent observation in both pigs and poultry but the increased fat deposition in poultry is an additional complication that will have to be resolved.

Phosphorus excretion

The environmental nutrient load is largely influenced by nitrogen and phosphorus excretion. Phosphorus excretion is particularly important in that it is implicated in the eutrophication of fresh waters and therefore is intimately related to the ecosystem that will prevail in bodies of fresh water.

Feed ingredients of plant origin are fairly rich in total phosphorus content but most of this is in the form of phytic acid which cannot be digested by monogastric animals. The digestible phosphorus concentration in plant feed materials varies from 15-45% of the total phosphorus and often a working value of 30% is used in feed formulations. By comparison the digestibility of phosphorus in meat meal is 75-80%, and in monocalcium phosphate is about 80%. Overall phosphorus retention by animals is however quite low in many standard feeds, and sows for example, only retain about 25% of the ingested phosphorus (van der Peet-Schwering and den Hartog, 2000).

The enzyme phytase, which hydrolyses phytic acid, has been very successful in reducing phosphorus levels in feeds and this in turn means less phosphorus excreted in manure (see Chapter 4). The principle is to feed animals with the appropriate level of phosphorus for optimal growth and performance while reducing the excretion on non-digestible phosphorus in the form of phytic acid. This requires formulation of low phosphorus diets with increased availability of endogenous phosphorus reserves in feed ingredients. The use of phytase enables monogastric animals to digest the phytate phosphorus contained in feed ingredients and this enables the phosphorus content of feed to be reduced.

Numerous trials and experiments have demonstrated that reducing the incorporation of inorganic phosphorus into feed and increasing the availability of plant phosphorus by the incorporation of phytase into pig and poultry feeds enables a large reduction in the excretion of phosphorus.

Phytic acid is also involved in the chelation of other minerals such as calcium, iron, magnesium, potassium and zinc. Use of phytase in feed formulations may also reduce the excretion of polluting minerals into the environment.

Other nutritional strategies such as the use of phosphorus supplements with high biological availability, and adequate amounts of vitamin D also need to be considered (Waldroup, 1999).

Odour control from livestock production and dietary manipulation

Reductions in the ENL are a very important aspect of environmental management. In the animal production industries however it is also extremely important to reduce the concentration of odorous emissions produced by livestock and by decomposition of manure. The fermentative activity of the microflora in the gastrointestinal tract of an animal produce many different molecules including ammonia, hydrogen sulphide, phenols, indole and skatole all of which have unpleasant odours and contribute to odour emissions.

Ammonia volatilazation from pig slurry should certainly be minimized to improve air quality inside the pig house as a welfare issue and to prevent high emissions into the environment. Ammonia is mainly formed from the enzymatic breakdown of urea in urine through the action of the urease enzyme present in faeces. The urea concentration of the urine and the pH of the slurry are important factors influencing the ammonia emission. Storage and distribution of manure also generate serious odour problems. This is a very important topic in the modern world and many strategies need to be explored including physical methods of manure storage, handling and disposal (Powers, 1999).

Manipulation of livestock diets to influence the microbial population in the gastrointestinal tract will alter the excreta composition and thus the odour of the excretions. This is certainly one important method that is effective in reducing odour from animal production units and many strategies are have been explored in attempts to manage the microflora as described in Chapter 5. Incorporation of fermentable carbohydrates into the diet, increased use of crystalline amino acids, increased feed copper contents and reduced dietary protein levels have beneficial effects upon odour emission in pig production. Use of organic acids, tea polyphenols, and extracts of the plant *Yucca schidigera* have all been investigated as potential odour reducing agents (Sutton *et al.*, 1999).

Feed raw material selection may influence manure odour, whether freshly excreted or during storage. Increasing amounts of blood meal in piglet feeds is associated with increasing odour intensity of the manure. Other research has shown that addition of peppermint to cattle diets improved excreted manure odour (Kellems *et al.*, 1979).

Reducing the dietary nitrogen intake can lower the urea excretion in pigs, and this in turn will reduce ammonia emissions. A 1% reduction of dietary protein level results in a reduction of ammonia emission by 10% on average. Low protein diets not only affect ammonia emissions however, but also emission of other odorous components such as hydrogen sulphide (Hobbs *et al.*, 1996).

The addition of various calcium salts of organic acids, particularly calcium benzoate, was effective in reducing both pH of pig slurry and ammonia emissions (Aarnink and Canh, 1999). Similarly the incorporation of adipic acid at 1% into piglet feeds reduced urinary pH from 7.7 to 5.5 and had a significant effect upon reducing ammonia emissions (van Kempen, 2001). Unfortunately in the EU neither adipic nor benzoic acids are permitted for use in feed so this strategy cannot be utilised at present.

High dietary copper content in pig feeds also significantly reduces the nuisance level of manure (Armstrong *et al.*, 2000). The odour and irritation intensity of faeces were lower from pigs consuming diets containing 225 ppm copper as $CuSO_4$ or 66 and 100 ppm copper as Cu citrate. Organic sources of copper seem to give improved odour characteristics of manure at lower dose levels than an inorganic copper source. Copper of course is a serious environmental pollutant itself and is already strictly controlled in animal feeds. Therefore feeding high levels of copper to reduce odour problems will solve one problem but create an even bigger one. Nevertheless copper nutrition may be worthy of further study in terms of its relationship with quality of manure.

Some research suggests that adding non-digestible oligosaccharides or various carbohydrates that are fermented in the large intestine to the diet improves nutrient utilisation by providing a more suitable energy:protein ratio for protein digestion. Often when formulating animal feeds the energy:protein ratio is considered. However perhaps in future more attention should be paid to the components of the energy source that will aid in odour prevention and reduce nutrient excretion.

There is increasing evidence that nitrogen excretion is shifted from urea in urine to bacterial protein in faeces when highly fermentable fibrous feedstuffs are included in the diet. More of the dietary nitrogen is used for microbial protein synthesis in the hind gut of pigs and is finally excreted in the faeces. Furthermore fermentable fibre can also decrease the pH of slurry by conversion into volatile fatty acids (VFAs).

Volatile fatty acids, important in controlling the pH of slurry, are mainly produced by bacterial fermentation, by deamination of amino acids in the hind gut of pigs and by anaerobic digestion of slurry during storage. Faecal VFAs are mainly produced by microbial fermentation of dietary fibre and only small amounts of VFA are derived from the urine (Sutton *et al.*, 1999). The rates of VFA production depend on the amount and characteristics of fermentable fibre. Complex carbohydrate sources such as sugar beet pulp, soybean hulls, coconut meal, cellulose and various non-digestible oligosaccharides tend to promote VFA production whereas lignin has a negative effect.

A reduction of ammonia emission from the slurry of pigs given dried sugar beet pulp has been observed (Canh *et al.*, 1998). Sugar beet pulp contains a considerable amount of fermentable fibre which can be used as an energy source for pigs. The sugar beet pulp is easily fermented in the hind gut of pigs because of its low lignin content and considerable level of pectin.

Adding up to 150 g/kg sugar beet pulp silage to pig diets which increased the amount of fermentable fibre to 365 g/kg did not influence the performance of growing-finishing pigs. The digestibility of amino acids was not depressed by sugar beet pulp and so overall nitrogen retention was not affected by the sugar beet fibre.

Sugar beet pulp however, had a marked effect upon the pH of the slurry. The pH of the slurry was lower in pigs fed a high amount of fermentable fibre compared to pigs on a low fermentable fibre diet. The quality of fermentable fibre in sugar beet pulp is ideal for bacterial fermentation with low lignin content and high pectin, cellulose and hemicellulose contents.

When fermentable fibre was increased to 365 g/kg by including 150 g/kg of sugar beet pulp large quantities of VFAs were produced in the faeces of pigs and in slurry during storage (Table 6). A high fermentable fibre content in the diet enhanced VFA production. In general as dietary level of fermentable fibre increased the amount of individual VFAs increased. Acetic acid was predominant in the pool followed by smaller amounts of propionic acid and butyric acid. The increased amount of VFAs resulted in a lower pH of the slurry and this persisted during storage of the slurry.

The ammonia emission from the slurry was also strongly affected by the level of fermentable fibre in the diet. The ammonia content, temperature and pH of the slurry are important factors influencing ammonia emissions. Inclusion of sugar beet pulp did not affect the amount of ammonia in the slurry as shown in Table 6 but had a marked effect upon actual ammonia emissions from the slurry. This was due to the increased amount of VFAs in the slurry. When tapioca was replaced by sugar beet pulp there was a clear proportionate reduction in ammonia emissions.

Increasing the level of fermentable fibre in the diet of growing-finishing pigs may be an economically feasible way to increase the concentration of VFAs in the slurry resulting in a lower pH of the slurry. This consequently reduces the ammonia emission from the slurry and reduces overall ammonia emission from pig production systems.

Table 6 Nitrogen and volatile fatty acid (VFA) content of slurry produced by growing-finishing pigs fed different levels of sugar beet pulp silage

Slurry Components (g/kg)	Sugar beet pulp silage in feed (g/kg)			
	0	50	100	150
Total nitrogen	6.28	6.71	6.52	7.15
NH_4^+ nitrogen	3.57	3.88	3.78	3.92
Acetic acid	5.11[a]*	9.15[b]	12.82[c]	14.18[d]
Propionic acid	1.80[a]	1.61[a]	2.28[a]	3.40[b]
Butyric acid	0.77[a]	1.67[ac]	2.45[bc]	3.28[b]
Total VFAs	9.04[a]	13.92[b]	19.15[c]	22.42[d]

*Values in the same row with different superscipts indicates $P<0.05$

Dietary manipulation is an important means to reduce manure odours before excretion and during manure storage when anaerobic decomposition is occurring and odour compounds are being formed.

Integrated pollution prevention and control (IPPC)

The IPPC is a directive of the EU aimed at providing a high level of protection for the environment. It covers not only agriculture but also the whole spectrum of industries which may contribute to environmental pollution. The integration of the control of pollution into air, water and soil is a balanced approach and covers a wide range of pollutants as shown in Table 7.

Nitrate levels in water
Ammonia emissions
Biochemical oxygen demand (BOD) pollutants of waters
Methane
Odour
Noise
Heat
Energy use

Table 7
Pollutants covered by the IPPC Directive of the EU

In animal production the IPPC only applies to pig and poultry sectors and not to the beef and dairy industry. Apparently pig and poultry are considered as intensive industries whereas beef and dairy are considered as extensive industries.

In general the IPPC programme requires the adoption of practices known as best available techniques (BAT). The actual mechanism of BAT in pig and poultry feeding for example are that all diets should be formulated to minimise excess protein fed and at the same time provide an adequate balance of amino acids. Diets should also be formulated to reduce phosphorus excretions. As indicated above it is possible to make considerable reductions in nitrogen and phosphorus excretions by careful feed formulation which should comply with the spirit of the IPPC system.

Future directions

Food and feed safety

Quality assurance systems such as GMP, HACCP or ISO 9000 will be more widely introduced into the feed manufacturing industry. A GMP system for animal feeds was introduced in the Netherlands in 1992 and this is now being extended towards suppliers of raw materials (Tielen and den Hartog, 2001). Such an approach will most likely become more widely practised in the EU. This will bring advantages of increased transparency and traceability into the feed manufacturing process which should help to improve the inherent quality of feed and to assure the human consumers that feed quality is closely monitored.

Animal welfare

Welfare issues will become ever more significant in the developed countries. It is partly related to the decreasing proportion of the population of these societies that have any close connection to animal agriculture and to the widespread availability of cheap food. Also modern society tends to see animals as surrogate human beings and to assume they have the same behavioural patterns, needs and desires as humans. Nevertheless there is a general consensus that animal welfare is an important issue and should be of a high standard.

Welfare is a difficult concept to translate into practical conditions of animal production and is difficult to assess objectively in an animal population. There will be an increasing need to establish the true perceptions and welfare status of animals. Also it will be important to consider welfare in terms of food safety. It may be questioned as to whether eggs produced from outdoor layers will be more or less safe than those produced by layers in cages.

The modern consumer in general prefers that food animals should not be treated with drugs and medicines. However when animals are suffering from a disease then clearly medication becomes a welfare issue. Total Nutrition can play a useful role here if it can maintain health and avoid disease without reliance upon drugs and medicines.

Research on acute phase proteins perhaps offers the best way forward in terms of having an objective measurement of welfare. It should be possible to relate nutritional, health and welfare status to levels of acute phase proteins. This is certainly an area of active research and levels of acute phase proteins could form the basis of a scientific assessment of welfare.

Utilization of feed resources

It is a remarkable and grossly uneconomic aspect of animal nutrition that pigs and poultry excrete some 65% of their total nitrogen intake and as much as 70% of the phosphorus intake. The unfortunate economic fact is that these nutrients have to be purchased, processed, distributed and fed to animals only to be excreted back into the environment. Clearly nutritional research in the future must focus upon digestion and absorption of nutrients. As indicated in Chapter 4 enzyme and phospholipid nutricines have already given some benefits here. However much remains to be done both in terms of enzymatic processing of feed raw materials to make them more digestible and in terms of improving digestion and absorption *in vivo*.

This is also an important ethical issue as it relates to the amount of food of animal origin that can be obtained from any given quantity of basic feed raw materials. Increasing population growth and demands for meat milk and eggs will require ever-greater efficiencies in the conversion of basic feed raw materials into high-value human food products. There have clearly been major and significant advances in animal nutrition over the last century. Nevertheless, a production system, which converts less than 50% of ingested nutrients into human food products, offers an enormous potential for further improvement.

Minimum crude protein in diets

Reductions in crude protein levels of the feed will certainly minimise nitrogen excretion and consequently ammonia emissions. This suggests that diets should be formulated to minimise crude protein contents with perhaps more attention paid to the ideal protein concept. There will be a need to ascertain the levels of dietary crude protein that will also contain adequate levels of both essential and non-essential amino acids for commercial strains of high lean gain pigs and fast growing broilers

However reduced feed protein content may also compromise growth performance of the animals, particularly in broilers. Consequently there is an urgent need to study further the limitations to nitrogen utilisation by animals in order to increase the proportion of nitrogen retained in the body and maintain good performance.

Several characteristics of Total Nutrition will be important here: application of feed enzymes, organic acid nutricines and management of the microflora in the gastrointestinal tract will all play an important role in nitrogen nutrition of animals.

The presence or absence of disease also has an effect upon nitrogen retention and also therefore upon nitrogen excretion. Feeding animals for health and growth will have a significant impact upon the environment as well as upon the economics of animal production.

Mineral nutrition

High feed copper has already been shown to favourably reduce the nuisance level of pig faeces (Armstrong *et al.*, 2000) and further research upon the precise effect copper exerts on manure quality could be worthwhile.

Another possible strategy to improve utilisation of phosphorus in feeds is to reduce the phytic acid content of cereal grains by plant breeding techniques. Low phytic acid maize seems also to have higher levels of available phosphorus than standard maize. Normal yellow dent maize had a total P of 0.23% and non-phytate P of 0.03%, compared to a low phytic acid maize with total P of 0.27% and non-phytate P of 0.17% (Waldroup *et al.*, 2000). Feeding low phytic acid maize in poultry diets was able to reduce phosphorus excretion in the manure (Li *et al.*, 2000). There is certainly much to be done in terms of genetically improving the quality of feed raw materials for environmental concerns as well as for nutritional concerns.

Endogenous losses

Endogenous losses of cells, recycled nutrients, and metabolites, contribute to the ENL. The endogenous losses are a function of the physiological activity of the animal and it is not easy to control this by nutritional or management techniques but nevertheless needs to be addressed to reduce further the environmental impact of animal production.

Management of animal waste products

Another major challenge for the future of animal production will be to manage effectively and sustainably waste products from animal production. This includes manure, and waste products from slaughter and processing of animals. This also involves issues of air pollution, water quality, soil quality, fly populations, and dissemination of pathogenic organisms.

Animal waste management research has largely been in the domain of agricultural engineers. It is technically feasible to ferment manure into methane which can be used as a fuel. Fairly dry poultry manure is readily incinerated or can be used directly as a fuel in power generation. However management of animal waste products is now also a topic of vital importance to animal nutritionists as recent events related to BSE and the prohibition of the use of meat and bone meals has illustrated. The situation now is that a valuable nutritional resource can no longer be utilized in animal feeds. Nevertheless the waste products from abattoirs still continue to be generated and must be disposed of in more costly and less efficient methods.

Dietary manipulation is a potential means of reducing the quantity of manure excreted, the unpleasant odours before excretion and during manure storage when anaerobic decomposition takes place. Considerably more research effort will need to be directed to establish which feed formulations and feed ingredients benefit both animal production and the environment.

General conclusions

The modern consumer in the developed world today has access to unlimited quantities of low cost food of very high nutritional and hygienic standards. Nevertheless there is considerable consumer unease about food quality and in recent years there has been a very significant public focus on feed manufacturing and systems of animal production. This has resulted in new legislation concerning animal welfare and nutrition. It has also encouraged the major retail groups to become much more actively involved in monitoring and controlling the production of the food products they purchase. It has culminated in a major publication by the EU of a white paper on food safety in 2000 which explicitly linked feed safety to food safety.

The concept of Total Nutrition is a possible response to these new demands where all the different steps in the animal feed chain is considered from quality of raw materials to utilization of feed by the animals. A judicious application of nutrients and nutricines in Total Nutrition should allow the production of safe animal feeds which will give good animal performance and help maintain the profitability of the industry in an acceptable manner.

The objectives in Total Nutrition of using nutrients and nutricines to support disease avoidance and health maintenance, to minimise environmental pollution, and to improve the welfare status of animals are very much in tune with modern consumer demands. Most nutrients and nutricines are of natural origin and this should help to improve the general perception of modern animal agriculture and go some way to alleviate consumer concerns. It will be important to emphasis that Total Nutrition addresses issues of growth, health, environmental pollution and the welfare status of animals.

References

Aarnink, A. J. A. and Canh, T. T. (1999). Ammonia emission from pig houses as affected by dietary composition. *Feed Mix*, **7**: 23-27.

Aletor, V. A., Hamid, I., Niess, E. and Pfeffer, E. (2000). Low-protein amino acid-supplemented diets in broiler chickens: effects on performance, carcass characteristics, whole-body composition and efficiencies of nutrient utilization. *Journal of the Science of Food and Agriculture*, **80**: 547-554.

Armstrong, T. A., Williams, C. M., Spears, J. W. and Schiffman, S. S. (2000). High dietary copper improves odor characteristics of swine waste. *Journal of Animal Science*, **78**: 859-864.

Bossman, D. (2001). Developments affecting the feed industry- an international perspective. International Feed Industry Federation Roadshow, Sun City, South Africa.

Broom, D. M. (1996). A review of animal welfare measurement in pigs. *Pig News and Information*, **17**: 109N-114N.

Byrd, J. A., Hargis, B. M., Caldwell, D. J., Bailey, R. H., Herron, K. L., McReynolds, J. L., Brewer, R. L., Anderson, R. C., Bischoff, K. M., Callaway, T. R. and Kubena, L. F. (2001). Effect of lactic acid administration in the drinking water during preslaughter feed withdrawal on Salmonella and Campylobacter contamination of broilers. *Poultry Science*, **80**: 278-283.

Canh, T. T., Schrama, J. W., Aarnink, A. J. A., Verstegen, M. W. A., van't Klooster, C. E. and Heetkamp, M. J. W. (1998). Effect of dietary fermentable fibre from pressed sugar-beet pulp silage on ammonia emission from slurry of growing-finishing pigs. *Animal Science*, **67**: 583-590.

COBB (1999). Commercial literature: Cobb 500 Maintaining the momentum

Commission of the European Communities (2000). White paper on food safety.

Flock, D. K. (1998). Genetic-economic aspects of feed efficiency in laying hens. *World's Poultry Science Journal*, **54**: 225-239.

Grote, H. and Radewahn, P. (2000). Future challenges facing the compound feed industry. *Kraftfutter*, **9**: 315-320.

Hobbs, P. J., Pain, B. F., Misselbrooke, T. H., Kay, R. M. and Lee, P. A. (1996). Gaseous emissions of slurry from pigs offered low crude protein diets. *Animal Science*, **62**: 635.

Jongbloed, A. W., Lenis, N. P. (1992). Alteration of nutrition as a means to reduce environmental pollution by pigs. *Livestock Production Science*, **31**: 75-94.

Kellems, R. O., Miner, J. R. and Church, D. C. (1979). Effect of ration, waste composition and length of storage on the volatilization of ammonia, hydrogen sulfide and odors from cattle waste. *Journal of Animal Science*, **48**: 436-445.

Ketels, E. (1999). Balancing amino acids to decrease nitrogen pollution. *Feed Mix*, **7**: 17-21.

Khush, G. S. (2001). Challenges for meeting the global food and nutrient needs in the new millennium. *Proceedings of the Nutrition Society*, **60**: 15-26.

Lee, P. (2001). Pollution regulations in relation to animal production in the UK. Society of Feed Technologist, January, pp 1-7.

Li, Y. C., Ledoux, D. R., Veum, T. L., Raboy, V. and Ertl, D. S. (2000). Effects of low phytic acid corn on phosphorus utilzation, performance, and bone mineralization in broiler chicks. *Poultry Science*, **79**: 1444-1450.

Millward, D. J. (1999). Meat or wheat for the next millennium? *Proceedings of the Nutrition Society*, **58**: 209-210.

Piñeiro, M., Alava, M. A., Lorenzo, E., Piñeiro, C. and Piñeiro, A. (2001). Characterisation of the acute phase protein response in pigs affected by transport related stress. The Second European

Colloqium on Acute Phase Proteins, University of Bonn, Germany. www.http://gla.ac.uk/faculties/vet/research/protein/index.html.

Powers, W. J. (1999). Odor control for livestock systems. *Journal of Animal Science 77*, Suppl. 2. *Journal of Dairy Science* **82**, suppl 2 : 169-176.

Rosegrant, M. W., Leach, N. and Gerpacio, R. V. (1999). Alternative futures for world cereal and meat consumption. *Proceedings of the Nutrition Society*, **58**: 219-234.

Sutton, A. L., Kephart, K. B., Verstegen, M. W. B., Canh, T.T. and Hobbs, P. J. (1999). Potential for reduction of odorous compounds in swine manure through diet manipulation. *Journal of Animal Science*, **77**: 430-439.

Tielen, M. J. M. and den Hartog, J. (2001). Feed safety for high quality of food of animal origin. In: *Advances in Nutritional Technology 2001*, Edited by A. F. B. van der Poel, J. L. Váhl and R. P. Kwakkel. Wageningen Pers, The Netherlands, pp. 249-257.

Tuitoek, K., Young, L. G., de Lange, C. F. M. and Kerr, B. J. (1997). The effect of reducing excess dietary amino acids on growing-finishing pig performance: An evaluation of the ideal protein concept. *Journal of Animal Science*, **75**: 1575-1583.

Van der Peet-Schwering, C. M. C. and Den Hartog, L. A. (2000). Manipulation of pig diets to minimize the environmental impact of pig production in the Netherlands. *Pig News and Information*, **21**: 53N-58N.

Van Kempen, T. A. T. G. (2001). Dietary adipic acid reduces ammonia emission from swine excreta. *Journal of Animal Science*, **79**: 2412-2417.

Varel, V. H. and Yen, J. T. (1997). Microbial perspective on fiber utlization by swine. *Journal of Animal Science*, **75**: 2715-2722.

Waldroup, P. W. (1999). Nutritional approaches to reducing phosphorus excretion by poultry. *Poultry Science*, **78**: 683-691.

Waldroup, P. W., Kersey, J. H., Saleh, E. A., Fritts, C. A., Yan, F., Stilborn, H. L., Crum Jr., R. C. and Raboy, V. (2000). Nonphytate phosphorus requirement and phosphorus excretion of broiler chicks fed diets composed of normal or high available phosphate corn with and without microbial phytase. *Poultry Science*, **79**: 1451-1455.

Whyte, P., Collins, J. D., McGill, K., Monahan, C. and O'Mahony, H. (2001). The effect of transportation stress on excretion rates of Campylobacters in market-age broilers. *Poultry Science*, **80**: 817-820.

INDEX